Developments in Atmospheric Science, 3

Methods in Agricultural Meteorology

Developments in Atmospheric Science, 3

METHODS IN AGRICULTURAL METEOROLOGY

by

LIONEL P. SMITH

President of the Commission for Agricultural Meteorology
World Meteorological Organization, 1962-1971
Meteorological Office
Bracknell, Berks., Great Britain

ELSEVIER SCIENTIFIC PUBLISHING COMPANY
Amsterdam - Oxford - New York 1975

ELSEVIER SCIENTIFIC PUBLISHING COMPANY
335 JAN VAN GALENSTRAAT
P.O. BOX 211, AMSTERDAM, THE NETHERLANDS

AMERICAN ELSEVIER PUBLISHING COMPANY, INC.
52 VANDERBILT AVENUE
NEW YORK, NEW YORK 10017

Library of Congress Card Number: 74-21868

ISBN 0-444-41286-7

With 22 illustrations and 42 tables

Printed in The Netherlands

Preface

"Men argue, nature acts."
VOLTAIRE

This book attempts to explain the principles underlying agricultural meteorology, and show some of the methods by which progress can be made.

Most of the examples quoted in the text refer to work with which the author was associated. This does not imply that they are the only, or even the best, examples. The Appendices contain advice on the way in which the many papers on the subject can be consulted.

The first duty of a scientist is not to worship knowledge, but to question it. The following pages are not intended to provide a set of answers. They are written for the applied research worker who is willing to pit his wits against those who are armed only with dogma and ignorance, and pose the question, "Does this idea help you?"

If the answer is "Yes", then the author's aim will have been achieved.

The author wishes to convey his thanks to: Mr. Robert Graves, for permission to quote from his works; the Secretary-General of the World Meteorological Organization, Geneva, for permission to reproduce certain diagrams; his colleagues and collaborators over the years who have allowed him to refer to some of the results of their various investigations; the Director-General of the United Kingdom Meteorological Office for permission to publish this quintessence of some quarter of a century's work; and his many friends and aquaintances in almost every country in the world, with whom he has spent many rewarding hours of discussion and from whom he has learnt so much.

Figs. 11, 12 and 13 are the copyright of the World Meteorological Organization and are reproduced by permission of the Secretary-General.

Contents

CHAPTER 1

The Meaning of Agricultural Meteorology

"Fortunatus et ille deos qui novit agrestis."
VERGIL

THE INTENT

Agricultural meteorology aims to put the science of meteorology to the service of agriculture in all its various forms and facets, to improve the sensible use of land, to help to produce the maximum food for humanity and to avoid the irreversible abuse of land resources.

Numbers are, or should be, the language of science, and the use of numbers enables an art to be transformed into an applied science so that skills can be learnt more quickly and more efficiently, and errors or disasters can be avoided. There is nothing quite so frightening as the sight of ignorance in action.

The pursuit and establishment of full scientific truth is part of this process, but such knowledge is sterile unless it is put to use, and unless it is made available in a way in which others can use it. The increase of basic knowledge depends to a large extent on the input of effort, involving brains, time, money and facilities. As a general rule, the more intensive the effort, the more successful the outcome, but many a research programme, conscientiously carried out, leads to frustrating minimal discoveries. Except in the rare cases of genius, or fortunate serendipity, progress is distressingly slow.

Any urgent practical problem demands an immediate answer, and such slow progress cannot be tolerated. Furthermore, a given problem demands a reasonable minimum of reliability and yet cannot afford a complicated expensive infra-structure of effort. The applied scientist has therefore to simplify a solution as far as possible but in so doing, sacrifice the minimum of accuracy. The more basic knowledge that is available, then the easier becomes this simplification process. If knowledge is limited, then empirical or semi-empirical solutions have perforce to be adopted.

Thus although pure research and applied research appear to work in opposite directions, one complicating and the other simplifying, they are in fact working together, each making use of the progress made by the other. The research scientist tends to try to answer questions which he puts to himself; the applied scientist has to provide some sort of an answer to questions raised by others. Both work partly as individuals, partly as members of teams. The applied scientist, in particular, has to realize the importance of interdisciplinary work. His is no ivory tower, but a structure of far more plebian material, with many doors opening to the minds and ideas of associated sciences. Without continual consultation and cross-fertilization he can make but little progress.

Collaboration and co-operation are therefore essential, if only for the fact that it is very difficult to learn sufficient about more than one or two scientific disciplines. The agricultural meteorologist, preferably first gaining his experience in the ubiquitous science of meteorology, has to learn to listen to the details pertaining to the associated science, he has to learn to appreciate their significance in relation to the physical conditions of the environment, and he must also be able to think clearly on his own, thinking both outwards and inwards, extrovert and introvert. He must retain some reverence for the past and recognize the inherent virtues of old methods, but must be enough of an iconoclast to be able to reject false reasoning. He must be cautious in the advocation of new ideas, not giving tongue before he is tolerably certain of his new evidence and confident in his powers of reasoning, and yet he must have sufficient courage to put forward an interim answer. Somehow or other, he aims at the golden mean between the fool that rushes in and the angel who fears to tread.

Agricultural meteorology is not easy to teach or to impart to others, because success in the subject depends largely on intrinsic ability, basic character, and acquired experience. The work is similar to that of a detective engaged on criminal investigations, who has to be able to identify the significant clues and make logical deductions from a mass of apparently conflicting evidence.

It is sometimes difficult to find any clues at all, and yet at other times there may be a profusion of apparently relevant factors. On such occasions it is most important to be able to recognize which of such factors are of first order importance and which are only of a secondary influence. It is relatively easy to complicate a problem and to strive after sometimes illusory accuracy by over-attention to unimportant details, but the successful applied scientist must have the ability to identify and quantify the really important factors appropriate to the situation.

Success demands the insight of a poet who, while appreciating the presence of contributory causes, has the instinct to recognize the underlying dominant reasons. It was Hilaire Belloc who, in his older years, recalled a romance of his youth — "I forget the name of the girl, I forget the name of the village, but the wine — it was Chambertin."

THE EXTENT

Because the physical weather conditions play such an intricate part in problems concerning land use and food production, it is not easy to present a clear concise summary of the extent of agricultural meteorology. Certain major areas of concern can, however, be specified in the following manner.

Plant physiology

Under this heading is considered the "development" and "growth" of plants of all sizes from lichen to trees, The word "development" implies the phenological stages of crop growth, as for example, the change from leaf production to seed head formation in a cereal, or the emergence of buds in a fruit tree. Development therefore deals with the beginning and end of a type of growth; growth, in this context, is concerned with the extent of photosynthesis and translocation of material within a plant, during each growth stage.

The three major features of development, without which the crop cycle cannot be fully completed, are germination, pollination and maturity, all of which are sensitive to weather conditions. Nevertheless, it is true to say that plants must be able to complete all their necessary stages of growth in order to survive, so that the weather must be such as to permit them to do so within the time available. Normally this available time lies within a calendar year, but this is not invariably so. Some plants need two years of suitable weather in order to complete their full cycle. Seeds have been known to lie dormant, and yet remain viable, over long periods of years.

The feature known as "growth" deals principally with the plant factors which affect the quantity and quality of the end product, namely the yield. These final yields are in many cases dependent on satisfactory growth during earlier stages of crop development, and therefore growth and yield are not always identical in meaning. The quality and provenance of the initial seed are also important, implying that conditions prevailing in previous crop cycles have a bearing on the problem.

Crop yields vary in type; they may be concerned simply with green matter production as in a forage crop, or in a vegetable such as a lettuce or cabbage; the end product may be the root of the crop, providing food for animals or mankind, or it may be the trunk of a tree, as in forestry; it could be only a relatively small portion of the plant such as a fruit or a cereal grain. Much activity in agricultural meteorology has concerned itself with the search for relationships between weather and yield, but such is the complexity of the processes involved that simple answers cannot be expected to give other than approximate results with strictly local applications. Such approximations are nevertheless of considerable value, for without adequate food man cannot survive, being as he is, a parasite on a vegetable, and efficient food production demands a working knowledge of the effects of weather.

No food crop is any use to man until it is harvested; the weather conditions immediately before, during, and immediately after harvest are thus of paramount importance. Adverse conditions at this critical period can easily outweigh a year of favourable weather for development and growth.

Animal physiology

Except in areas where local practice forbids the human consumption of the flesh of all or certain animals, most farm livestock are produced with the aim of providing meat, although there are also very important animal products such as milk, wool and leather. Alternatively, they may be bred to provide a form of transport or haulage.

In each case the ability to reproduce the species, the fecundity of the breed, is the factor of basic importance. The period of gestation may be long, the reproduction rate is often low, for only pigs of the usual farm animals produce more than a few offspring in each litter. The physical weather conditions at conception, during pregnancy and at birth are therefore of the highest significance, either directly in the form of environmental stress or indirectly through the quantity and quality of available food.

During the lifetime of the animal, the same two types of weather effect, direct and indirect, continue to influence such production factors as gain of live weight or milk production. It is probably true to say that the indirect effect of weather via the food, is the most important in this context, although in extreme conditions of heat or cold this may not be so. Because of the mobility of animals, and their conscious choice of action or feeding procedure, it is often difficult to specify the meteorological conditions which influence them, more so than in the case of a stationary plant. This difficulty is, of course, removed, when animals are constantly housed and hand- or machine-fed, but on a world scale, the vast majority of farm animals are free-ranging for most of the year.

Another complicating factor is that of management; a skilled and sympathetic milker, herdsman or shepherd can improve the performance of his animals in ways which cannot be explained in terms of physical or nutritional conditions; it has not yet been proved that plants respond in the same way to what can only be described as kindness.

Birds such as hens, ducks, geese and turkeys can also be included under this heading, both for their potential as food and for the eggs they produce. Many of the above remarks are equally applicable to this form of husbandry. The creation of controlled rearing and production conditions has probably made more progress in recent years in the realm of poultry, especially for edible chickens and eggs, than anywhere else, and once the environment becomes under the control of the farmer, then it is all the more important that correct knowledge is available of the optimum conditions, and the means whereby they can be provided. Despite the emotional reaction against what is called "factory farming" (itself an emotive choice of phrase and therefore scientifically suspect), there is little doubt that the increasing demand for food will demand greater operational efficiency.

Natural plants and wild animals

The two preceding paragraphs have considered only cultivated plants and domesticated birds and animals. With the awakening sense of responsibility towards the biological environment now making itself felt in many communities, the techniques of agricultural meteorology are being applied as part of an ecological approach to problems of conservation in general. True ecology cannot afford to omit due consideration of physical conditions, and indeed, unless it can include an accurate assessment of the influence of weather and climate, it is doomed to failure, or, at best, to incomplete success.

All the principles involved in the consideration of farm activities apply in regard to biological and zoological activity in the wild. The multiplicity of microclimates involved do, however, make a complex situation even more complicated. Much has been made of the influence of man's activities on natural habitats, especially in regard to their constant reduction in area, but even a small change in climate over a period of years, or a chance meteorological hazard such as an extreme winter or a lightning strike and consequent forest fire can have far-reaching immediate effects on an ecological balance.

It has been truly said that the only constancy in climate is its inconstant behaviour. Unless such continual variations are recognized and translated into their biological effects, we cannot hope to understand the problems of conservation. Meteorology in the service of agriculture, therefore, must move into the less controlled but equally urgent fields of meteorology in ecology. The weather is, as yet, an uncontrollable and often an unpredictable variable, but this does not mean that it can be ignored.

Attempts to influence the weather

Although external weather is largely outside the present firm control of man, the same is not true of inside weather. For example, in growth chambers or phytotrons, the weather is wholly artificial. It may not be possible to reproduce in this way the exact outside conditions, but at least many of the meteorological variables are under the dictate of the operator. The exact specification, in full physical terms, of this artificial climate needs to be fully understood before logical deductions can be made in terms of unqualified application. It is especially difficult to translate experimental results obtained under these conditions into the similar but essentially different conditions of the open field.

A lesser degree of control is obtained by the use of heated glass or plastic structure, wherein further modifications may be introduced such as additional light or carbon dioxide and controlled humidities. The external conditions obviously now have a considerable influence, and the degree to which such outside climate can be modified is strictly limited, if not by physical restrictions, at least by economic considerations.

An even lesser degree of control is available by the use of unheated plant cover by glass or plastic. Although the grower does in this way considerably alter his growing climate, he is far more dependent on the external weather.

Whatever the degree of control, the search for efficiency, on a practical scale, must be carried out with full consideration of the physical processes involved. The extent to which climate can be improved by the use of any form of crop cover depends on the climate of the location, the materials available and the degree of expertness used in their handling. Knowledge of the meteorology of protected cropping is essential for success, especially when the large outlay of capital per unit area is considered. Similar considerations arise in regard to any other form of agricultural building, whether it be a controlled atmosphere fruit store or a semi-open barn for the storage of hay. The housing of animals is another large area of research in which the meteorologist has a part to play, and again the scope varies from intensive poultry housing to a temporary structure for lambing ewes.

The losses of harvested crops during storage, either directly or indirectly through the onset of noxious pests or fungal diseases can be considerable, and in each case the internal climate of the storage conditions plays a dominating part. Similarly, conditions during transport to market or consumer can adversely affect the final useful yield of any enterprise. Losses due to frost or excessive heating are avoidable by correct interpretation of the weather risks and the modifications thereto provided by the means and condition of transport.

Long-term modifications of growth conditions

Apart from the use of structures, there are other ways in which the plant and animal climate can be modified on a long-term basis. One of the oldest of these is the practice of land drainage, which is designed to remove excess water from the soil, especially at times when it is harmful to crop growth or renders pastures unavailable for animal grazing.

The design of a drainage system is more a problem of soil physics than of meteorology, but the specification of the drainage need is certainly a question of the correct interpretation of the water balance, demanding an accurate assessment of the rainfall and evaporation climates which is best carried out by the agricultural meteorologist. Furthermore, the effect of drainage, and the extent to which it can decrease delays in cultivations and sowing, and hence avoid loss in yields, especially of spring-sown crops, is essentially one which demands the help of such a scientist.

Another long-term effect on crop climate is the provision of shelter from the wind. As decrease in wind strength, and hence the avoidance of stomatal closure is probably the most important effect of a shelter belt or wind break, the questions of air flow beyond such artificial or natural structures are of prime importance in shelter design. There are also other meteorological effects needing consideration, such as those pertaining to shade and shadows, alterations of temperature and humidity, possible changes in evaporation, and secondary effects such as the deposition of pollutants or pests and the incidence of plant disease. Good shelter must be carefully planned, and careful planning demands full meteorological knowledge.

Although shelter from the wind may be a worthwhile design feature in windy climates, in countries with a strong overhead sunshine shade may be more important. Some crops, such as coffee, need such shade and many animals such as dairy cows thrive better under conditions which reduce an excessive heat load. Again the problem is one which needs meteorological help for its solution.

Short-term modifications

The oldest and most obvious illustration of short-term alterations of growing climate is the use of irrigation, whereby deficiencies in natural rainfall are made good by the addition of water to the soil in appropriate amounts at correct intervals. The calculation of irrigation need and the devising of irrigation schedules based on meteorological observations or calculations have made immense progress in recent decades, and few major schemes are now planned or carried out without due reference to the weather factors.

So much has been done in regard to soil moisture control that the meteorological aspects of problems of soil nutrient control have tended to be overlooked. The addition of fertilizers to the soil, whether natural in the form of manure or artificial in the form of chemicals is now a well-established practice in advanced agricultural countries, but it is also beginning to be realized that the timing of applications and the optimum quantities of fertilizer or lime that are needed are also functions of the seasonal weather.

A further example of short-term local modification is the use of soil mulches of various kinds. The alteration of the nature of the soil—air interface causes considerable repercussions on both the climate of the soil and also the conditions in the air immediately above the mulch and surrounding the growing plants. Once again, the effects can be direct, as those on temperature, soil moisture, humidity, evaporation, or they can be secondary through the consequent effects on soil bacteria, soilborne pests, water infiltration, gas exchange and so on. The cultivation of the soil in ridges, the choice between deep and shallow ploughing, the use of fallow and many other soil management techniques also involve meteorological considerations and physical interpretation.

Combatting adverse effects of weather

One of the more obvious adverse effects of weather is the occurrence of frost, which provides a threat to successful agriculture in many parts of the world. Without doubt the best way to combat frost is to avoid it, that is to say, to select areas where frost is least likely, or to so arrange the cropping programme that the danger of frost incidence is reduced to a minimum, a problem which will be returned to later when considering the field of agro-climatology.

There exist many methods which attempt to protect crops against frost, which attain varying degrees of success. The extent to which such methods are needed, the times when they should be brought into operation and the assessment of their efficiency are all problems of agro-meteorology. It is particularly important that reliable advice is made available to farmers and growers in this matter, because otherwise much money, time and effort will be wasted, and the potential losses due to frost are appreciable on a national scale and can be disastrous to the individual grower.

A more universal danger than frost, both in time and place, lies in a secondary effect of weather, namely the incidence and intensity of pests and disease. There is no crop which is not subject to such attacks, often of many different kinds, which reduce the productivity of all forms of agriculture, and in every case there is a weather factor which needs to be evaluated. Potentially, this is the type of problem in which agro-meteorology can play its most important economic part in regard to seasonal activities and annual production. The reason for this is that if meteorology and its applications can help to identify or predict the onset of a pest or disease, action can then be taken in the most efficient way to combat and restrain the attack by timely appropriate action on the part of the cultivator.

Foreknowledge of an event, against which there is no possible remedial or protective procedure, serves little use, but merely extends the period of anxiety. When, however, a warning can be given and time and techniques are available to reduce the adverse effects, then the basic requirements of a valuable form of service of applied science have been met. Concentration of the efforts of agricultural meteorologists in this type of activity can result in gains in primary production out of all proportion to the cost of research and subsequent advisory service.

As in plant protection, similar problems exist in regard to helping to reduce losses of animals or animal production due to pests or diseases. In several cases the difficulties of providing such help are eased because the time lag between the significant weather and the subsequent effects are greater than in the case of plants. This permits advice to be given and action to be taken without the need for rapid assessment of a danger using only limited evidence, and without an expensive requirement for instant communications.

The threat of pest or disease losses is so great that preventive or curative action in the form of dosing or spraying is often taken on a routine basis, based on calendar or phenological timing. This may be, and often is, effective, but it is not always economic and it opens the door to possible pollution effects. The only sensible form of protection is the minimum effective method, and such a method cannot be established without skilled interpretation of the meteorological factors.

It is also pertinent to suggest that most chemical methods have within themselves the seeds of their own impotence, or rather a self-induced sterility. The constant kill of (say) 99% of a pathogen will encourage the growth of the 1% resistant strain, so that this in time will become dominant and so the protective method becomes less and less effective. This is a battle which is never completely won, and which needs a full knowledge of causes, effects and side-effects in order to avoid disaster.

Avoidable dangers

Frost has been quoted as a danger which is best avoided rather than combatted. Choice of site is not always possible, but choice of action at a given site gives a better opportunity for sensible husbandry. This is particularly true in regard to the minimising of pollution risk, whether in the soil or in the air. Reference has been already made to the general use of biocides which, if inexpertly or over-enthusiastically applied, can lead to local or more widespread contamination of the environment.

Spray drift is a meteorological problem, both in regard to the efficiency of the spray itself and also in the possible escape of spray material downwind to other crops which it may damage. Controlled burning, whether of unused grassland, cereal stubble or heather can be dangerous without a degree of appreciation of the weather factors. Water pollution in lakes, streams or rivers due to agricultural effluents is a hydro-meteorological problem. Indeed the problems of water quality in general, whether concerned with irrigation potential or suitability for human or stock consumption, are to a greater or less extent concerned with the meteorological factors of the water balance and the hydrological cycle.

In general, however, the farming areas of the world are more polluted than polluting, in that they are often at the receiving end of air-borne or water-borne dangers. There is one major phenomenon, hwoever, for which the users of the land may be designated the prime culprits and that is erosion.

To acquire a sense of proportion in this matter, it is salutary to reflect that the half-life of, say, a radio-active pollutant, may be measured in days, but that the half-life of erosion is measured in centuries. Areas eroded as a result of man's ill-considered action two thousand years ago are still to-day sterile and unreclaimed. Even in modern times, mistakes are made which lead to a process of soil erosion which is virtually irreversible.

Erosion may be a natural process essential to soil formation, but it can be disastrously accelerated by man's stupidity or cupidity. The basic processes of such erosion, once man has provided the circumstances which bring them

into effect, are the actions of wind or rain, and the duty of an agrometeo-
rologist is to study such processes and understand their occurrence and
effect. This being done, it is then possible to identify areas at risk and to
specify those forms of land mis-use which would lead to a loss of valuable
irreplaceable topsoil. Again we have a major problem where the help of an
applied meteorologist is needed by the soil specialist and the agronomist.
Prevention is far better than cure, but when reclamation action is possible
and is undertaken, the meteorologist again has a part to play.

Unavoidable dangers

Reference has already been made to the air-borne transport of pollutants,
and any form of such contamination, whether of a pathogen, pest, or chemical, arriving at a site can be said to be relatively unavoidable. The farm or
field cannot be moved to avoid the onset of any particular current of air, or
any fall of rain which has collected and is depositing the pollutants from the
air above.

Almost the entire process of the air transport of visible or invisible pollutants is influenced or controlled by meteorological processes, a fact realized
by the growing importance of the relatively new science of aerobiology.
From take-off to landing, through dispersal and dilution, including survival,
the multi-causal process is concerned with meteorological actions and reactions both on the micro- and macro-scales.

Thanks to fundamental research work in atmospheric physics, and because
of the needs for forecasting the weather and the demands of aeronautics,
much is known of matters such as turbulence, convection, diffusion and
long-distance trajectories. The majority of this existing knowledge and available current data is still to be put into use in regard to aerobiology, although
considerable progress along these lines has been made in recent years. Air-
borne pollution cannot be stopped except at source, but the recipient is at
least entitled to know his degree of risk and the state of affairs which bring
about his liability to danger.

A more direct form of unavoidable danger to agriculture lies in the form
of extreme weather, such as hurricane winds, flooding of all types, snow-
storms, heavy damaging rain, hailstorms, thunderstorms and lightning. Often,
little enough can be done to avoid such catastrophes, but at least loss of
human and animal life can be minimized by adequate forewarning. Again,
the quantification of the degree of risk can be of assistance, more especially
in regard to long-range planning, rather than an aid to short-term action
which demands very accurate forecasting.

More help can be given in regard to another type of major danger, that
concerning fires in forest, bush or prairie, which follow periods of drought
and can often be started by a lightning strike. Most countries in the world
with large forests have schemes in operation which take into account the

appropriate meteorological factors. Little enough can be done to prevent the start of a fire, except perhaps the closing of the area to tourists and campers likely to light a fire for cooking or throw away a cigarette or lighted match, but the fire-fighting services can be alerted and efficiently directed with the aid of meteorological services.

Decision making

Ever since agriculture began, and since the priests of the ancient religions pronounced the times for sowing and harvest, the farmer has made his day-to-day decision with at least one eye on the weather. The traditional methods have stood the test of time, at least with partial success, if only for the fact that failure often implied starvation. Modern decision making still rightly retains what it considers good of the age-old wisdom, but it is turning more and more towards numerical solutions, and an art is slowly changing into a science.

A computerized farm may be the materialist's dream (and a humanist's nightmare) but any form of calculation is limited by the accuracy of its data and by the validity of its method or programme. The dangers of an over-optimistic and unthinking import of empiricism, or the excessive unjustified simplification of a complex interrelated set of processes can be horrifying, especially when they occur on a large scale. If decision making is left to the individual in a number of small enterprises, there is at least a chance that half will be right and half will be wrong. If the decisions are centralized or applied to large units of production, an error in calculation becomes a total error throughout.

With the increasing tendency towards larger agricultural units, the numerical assessment of the effects of weather is therefore of increasing importance at all levels of management. The correct estimation and interpretation of the temperature and moisture status of the soil with a view to deciding the best time for cultivations, sowing, fertilizer applications and so on is a major example of the way in which science is coming to the aid of visual appraisement geared to traditional practices. Meteorological assistance with drainage problems has already been mentioned, but any help that can be given to avoid damage to soil and to take the correct action to avoid errors of judgement can have far-reaching effects. An incorrect decision at and around sowing time may not be correctable during the lifetime of a crop, and damage to soil structure may take years to remedy.

Decisions made in the light of the existing state of affairs, or with knowledge of the effects of past weather, for example the leaching of lime or fertilizer due to previous excessive rain, should be relatively easy to help in a scientific numerical manner, provided that the processes are reasonably understood. It is when the decision has to be made with the future weather in mind that a degree of chance enters into the arena. Weather forecasts are not perfect, and the longer the range of time they refer to, the more imperfect they tend to become. Nevertheless, decisions have to be made; they cannot be avoided by undue awareness of the uncertainties of the future, so that at the very least, meteorological assistance should aim at the provision of some estimates of the degree of risk.

Such estimates can be climatically-based or memory-based relying on personal experience, but they can also be conditioned by current factors which seem to alter the degrees of chance. An example for the need of such estimates of risk could be the choice of a cereal variety, picking the one most suitable for the type of future weather which is expected. A further example would be in livestock management, deciding how much winter keep should be stored.

Continually, throughout the farming year, decisions have to be and are made which involve correct interpretation of past weather, and a degree of appraisal of probable future weather. Sometimes, the farmer has no option but to carry out his tasks whatever the weather, but his choice of timing is still critical and as far as possible luck should be eliminated from his mental or mechanical calculations by proper use of applied science.

Agro-climatology — the strategy of land use

There is little doubt that in certain countries there is a highly efficient state of affairs in regard to the sensible use of land for agricultural purposes. It is true that there are changes from time to time, induced by economic circumstances, or by minor fluctuations of climate, or by technical developments, but in the main they are in a good state of proved relationship with soil and climate.

There is also little doubt that major changes or development plans for the long-term strategic use of land involve heavy outlays of capital far greater than the operational costs in any one season. It is therefore all the more important that such changes or plans should be based on a sound appraisal of the climatic advantages and climatic risks. The science of agro-climatology aims at the interpretation of the climate or the expected variety of weather over a decade or so in terms of the most suitable form of crop production or animal husbandry.

If all the agro-meteorological relationships were known so that each facet of the weather was correctly linked to each effect it has on agricultural processes, then it would be theoretically possible, no matter how intricate and laborious, to interpret the climate or concensus of weather into terms of land use. Manifestly this is not so; therefore, a state of affairs arises whereby progress takes place at both levels. A summary of agricultural events is linked to the summary of weather, and the relationships so implied are subjected to check by agro-meteorological investigations. For example, comparison of the areas wherein a given crop is grown, and where the highest yields are obtained, with the climate of such areas can give useful information regarding the weather limitations of the crop and the optimum growing conditions. Conversely, reliable findings in one particular aspect of agro-meteorology can be used as a criterion when extrapolating crop—climate relationships and helping in the correct introduction of a crop into a new area.

A great deal of what can be termed classical agro-climatology has been presented in small-scale maps, so that whole countries or even continents are represented on a sheet of foolscap. Such work undoubtedly is useful in presenting the broad details of climate and land use, but far more intricate work is needed before it can be useful on the planning scale. In other words, the macro-picture must be filled in with the fine details of the meso-scale analysis.

Work of this nature needs a highly qualified and experienced type of practitioner. Experience has shown that it is often better to work with a team of scientists rather than expect one person to have all the expertise that is necessary. Certainly, an agro-climatic survey of a limited area seems to demand a minimum of two, one from an agricultural discipline, the other an agro-climatologist.

The problems will be considered in detail in a later section, but it is sufficient to say at present that agro-climatology or the application of meteorological knowledge to questions of future strategic action present the greatest challenge to the applied scientist. A breadth of knowledge and a depth of judgement are both needed in order to give the maximum help in a set of problems on the solution of which the future preservation and utilization of the inhabited world may well depend.

Summary

The extent of agricultural meteorology can be summarized by the following headings. The list of subjects mentioned is not exhaustive, but it does cover the majority of the important types of meteorological applications to agriculture:
The plant cycle; growth, development and yield
The animal cycle; growth, by-products and yield
Meteorological aspects of natural ecology
Artificial growing climates; crop cover by glass or plastic
Farm buildings, animal housing, food and fodder storage
Drainage
Shelter from the wind; shade from the sun
Irrigation: soil cover and mulching: frost liability and frost protection
Pests and diseases of crops and animals
Soil and air pollution
Erosion control
Meteorological hazards; storms and fires

Decision making within a given plan on tactical short-term time scale

Strategic decisions for long-term planning; land use; national and international problems of food production and environmental conservation

THE PROCESSES

Micro-meteorology, treated as a branch of experimental or theoretical physics, is a specialist subject, and although it forms an essential part of agricultural meteorology, it is generally dealt with by specialists at only a restricted number of research stations. This does not imply that it can be divorced from the mainstream activities in the subject, but rather that it is a parallel activity. All meso-scale phenomena are built up by micro-scale actions, so that no semi-empirical work can be trusted unless it is supported and confirmed by micro-meteorological knowledge.

It is therefore important that the agro-meteorologist working in the meso- or macro-scales should be aware of the micro-meteorological processes and of the progress that is being made in research of this type. The difficulty of scale-conversion is a great one, and important though an advance in micro-meteorology may be, it does not always follow that immediate practical advances in applied science are available. On the other hand, no assumption of crop—soil—air relationships can be made without fundamental justification at the basic process level.

Interdisciplinary aspects

Four main types of scientific discipline are involved, namely the atmospheric sciences and the soil sciences which are concerned with the physical and chemical environment, plus the plant sciences and animal sciences (including all their associated branches such as pathology, entomology, parasitology, etc.) which deal with the contents of the biosphere.

The situation in practice is that it is difficult, if not impossible, for an exponent in any one of these broad categories to work independently of workers in other disciplines. Similarly, if any classification of responsibility is sought, it is almost impossible to allocate a particular process to a precise discipline without running the risk of controversy. The principle of "territorial imperative" does not only apply to the nesting sites of birds; it can also at times confound and confuse the best intentions of enthusiastic scientists.

To avoid such conflict, let it be stipulated here and now that it does not matter who does the work: the one thing that does matter is that the work is done correctly. The World Meteorological Organization, in 1972, published a *Technical Note* (No. 119; Smith, 1972), in which an attempt was made to indicate the main interests of the several sciences. The succeeding paragraphs owe much to that analysis, which probably represents the present opinion of the leading world micro-meteorologists, but which lays no claim to infallibility.

The physical processes

The processes themselves are involved and interrelated, so that they rarely occur in isolation, but it is convenient for clarity of thought, to consider them under headings which deal with six main subjects of transfer, namely radiation, heat, water, momentum, carbon dioxide and a miscellaneous group of other transfer constituents in solid, liquid or gaseous form which are found in the biosphere. The concern of the atmospheric scientist with respect to each of the groups of processes will now be discussed in turn, with brief references to those processes which principally lie within the realm of other disciplines.

Radiation

The fundamental role of the radiation balance must never be under-estimated in any agro-meteorological problem, no matter how difficult it may be to enumerate or evaluate its constituents. The incoming energy from the sun is the major source of power, the ability to do work without which no process can continue. If the sun fails to rise, the earth will die. On the other hand, the out-going radiation from our planet provides a counter-balance without which we would be consumed by heat. It is the balance of conditions that is all-important, plus the fact that a small proportion of the available energy is used for photosynthesis. Agriculture, from the point of view of a selfish human, is the way in which the energy of solar radiation is converted into the fuel needed for man's survival, in other words, his food. In particular, energy supply must never be confused in the mind with temperature regime. Temperature may be a result of an energy exchange, it may even at times and with due caution be used as an indicator of available energy; it must never be regarded as its complete specification. Temperature is not heat.

The most immediate effect of incoming short-wave radiation and out-going long-wave radiation is thermal, that is to say, the temperatures that arise as a result of the radiation balance. The distribution of solar radiation determines the temperatures over the surface of the earth, with modifications caused by season and by land—ocean interaction and general topography. The absorption of radiation is the major determinant of the temperatures of plant tissues, animals large and small, and of the rates of evaporation and transpiration. The thermal effects of radiation can never therefore be ignored or underrated in any agro-meteorological problem.

A further effect of radiation may be termed photochemical, and pertains to the radiation in the visible part of the spectrum and in the infra-red range, referred to in many scientific papers as "light". The process of photosynthesis takes place when leaves are exposed to light; that of photo-morphogenesis, involving the development phases of a plant, is sometimes controlled by radiation of specific wavelengths.

Another control in many plant species, and also of bird and animal behaviour is dependent on the duration ratio of day to night, in other words, the photoperiodic effect. It will be realized that this is the simplest form of seasonal calendar available to biological material, most especially in areas away from the Equator.

Light can also have an adverse effect, in that exposure to (say) ultra-violet rays can reduce the viability of viruses, with consequent beneficial effects for mankind. It can also increase vitamin contents.

There is thus a major requirement from meteorology of greater knowledge and homogeneous data regarding the spatial and seasonal distribution of radiation, both in regard to quantity and quality. In recent decades there has been an increase in the measurements around the world of incoming radiation but recordings of the radiation balance are made at only a few research stations and observatories.

A subject of even greater relative ignorance is the temperature of material at the surface of the planet. The rate at which substances lose heat by radiation depends on their surface temperature, and this radiant emission can be measured by a thermopile radiometer which gives an integrated value over the surface area influencing the instrument response and which does not interfere with the surface during the process.

This radiation can be calculated from a measurement of surface temperature, but such measurements are by no means simple. In particular the leaf-surface temperature is difficult to measure properly by means of a thermocouple or thermistor. Knowledge of surface temperatures is thus a problem for both the meteorologist and the biologist.

The most important field of radiation research for the biological scientist is clearly the response of vegetation to radiation of various qualities and quantities. He is also concerned with the plant geometry, the aspect presented by a tree or plant, and the interception of heat and light through a plant stand. The optical properties of a leaf or stem are also important, especially in respect of their powers of absorption or reflection.

In a similar fashion the animal scientist is concerned with body geometry and with the optical properties of the animal surface. The response of animals to radiation, both in regard to mobility and choice of habitat, and ultimately to the part played by radiation in their growth and well-being is also a fundamental area of research.

The soil, like any other surface, both absorbs and reflects radiant heat, so that its optical properties as well as its emissivity are the concern of the soil scientist. The all-important thermal regime of the soil and the way in which it can be modified for the betterment of plant growth, depend a great deal on the correct interpretation of radiation processes.

Heat

Although radiation from the sun conveys energy to the earth, the resultant heat, itself a form of energy, manifests its presence by the temperature of the substances themselves. Temperature, by itself, has no dimensions; it may control the presence or absence of biological processes, or even the rate at which they can take place, but the energy required to perform the process determines the extent to which the process can proceed.

The principal concern of meteorology in regard to heat transfer is that of the convection process, although a certain amount of conduction of heat does take place within the atmosphere and especially at the soil—air interface. Convection may be free or forced, in that airflow over obstacles can create vertical movements of air and hence translocations of heat. Within the soil itself, heat is mainly transferred by conduction, although the movement of water through the soil can have considerable thermal effects. Knowledge of the distribution of heat in the soil, and its variations with depth, diurnally and seasonally, are of the greatest importance in agro-meteorological research.

The response of a plant to heat varies from crop to crop and from variety to variety within a plant species. The resulting temperature, determined by the heat balance, exercises important biophysical and biochemical controls both in plant physiology and plant pathology. Similarly the response to temperature of animal production and reproduction is a major concern of animal science. Temperature is probably the meteorological parameter which is more frequently encountered in research papers of a biological nature. It clearly plays an important role, but its essential difference in nature, as compared to heat or energy, must be always kept clearly in mind.

Water

The energy balance and the temperature regime, are not by themselves sufficient to ensure biological growth, otherwise the optimum regions for food production could be in the middle of deserts. A balance of input and output of water is also essential. There are two reasons for this, one is that a plant cannot easily take in nutrients unless it also takes in water through its roots; the second is that it is the evaporation at the leaf surface of a plant that keeps it at a correct working temperature, otherwise it would gain heat, shrivel and die.

In many areas of the world, adequate soil-moisture supplies are the major limiting factors of the growing season, the prevailing temperatures not being limiting for plant growth. The acceptable adequacy often falls between narrow limits. Too much water in the soil leads to waterlogging, anaerobic conditions, root rot, and few commercial crops (except rice) can be grown under such circumstances. Drainage is a palliative and not a cure, and even if good drainage removes the excess water, the constant leaching of nutrients presents a difficult fertilizer problem. Too little water in the soil leads to permanent wilting and plant death.

Water addition to the soil by rain (or snow which is subsequently melted), is a meteorological factor; its redistribution within the soil and its retention within reach of the plant roots is a soil factor. The water extraction from the soil occurs either by direct evaporation from the soil surface or by transpiration from leaves, and in both cases is largely determined by the energy available from direct solar radiation, but which is modified by other meteorological factors such as temperature, wind and humidity, and by biological factors within the plant itself.

Many influences of the water balance on problems of animal husbandry are the indirect ones of fodder production for animal consumption. Even so, domestic farm animals need quite large quantities of water for drinking and can die of thirst as well as hunger. Excess of rainfall on animals presents a different set of problems, principally concerned with an animal energy balance designed to keep it warm; the process of drying a wet animal fur, skin or hide can demand a great deal of body heat, and pregnant animals in particular do not thrive in constantly wet conditions.

The role of the atmospheric scientist within this field of research can be stated in a few words, but it needs several volumes to describe in detail. Briefly it may be said to be the distribution and intensity of the sources of water and the complementary sinks of expenditure of moisture within the biosphere; in other words, precipitation, evaporation and transpiration.

Probably the most important question that has continually faced mankind is, "Will it rain tomorrow?" If it does not rain on some tomorrow, plants will not grow, food cannot be produced and the result will be starvation. The range of associated scientific problems is large and diverse, from the cloud-physics research concerned with the creation of rain (or snow) to the extensive climatology of precipitation, its amount, intensity, frequency and reliability. There are many gaps in the chain of knowledge, and probably the most significant is that dealing with extrapolation in time. Changes in rainfall regime or rainfall climate have, before now, caused the downfall of civilizations, population migrations and wholesale changes of the pattern of land use. Not all the changes have been for the worse, as far as mankind is concerned, but changes will continue to occur and our relative inability to foretell them is a considerable handicap, especially in critical marginal areas.

On the other side of the water-balance sheet comes the even more involved and unsolved problem of the return of water to the atmosphere or to the sea, requiring knowledge of the evaporation processes from inland water, land and plant surfaces. The incoming rainfall has been measured with a fair degree of success, but the same cannot be said of evaporation measurements, either in accuracy or extent. Attempts to quantify each item of the water balance continue to be made, by physicists, agro-meteorologists and hydrologists, but it must be admitted that many areas of uncertainty still exist, and the subject as a whole must be approached with care and caution. While in the past it was often thought that vegetation exercised the major control over transpiration intensity, it has now been realized that the atmosphere itself, and especially the energy available for providing the necessary latent heat of evaporation, often assumes the dominant role. The contribution of physicists to the biological problems of water use by plants has probably been one of the most important advances of the 20th century.

Mention must also be made of dewfall, which has not only fascinated the lay mind by its apparent magic, but also has often been inadequately dealt with by the scientist. Research has shown that roughly half the dew which is said to "fall", does in fact rise from the soil, the other half coming from moisture in the air. The role played by dew in plant growth has, in the past, probably been over-estimated; on the other hand, there is no doubt but that it can have a considerable effect on disease-infection conditions. The distribution of the sources of dew, its occurrence and its intensity, together with the transfer mechanisms at the soil—air—plant interfaces are subjects which deserve further critical examination and study.

The pedological problems are concerned with the input of water into the soil, the infiltration and internal redistribution of soil moisture. In particular, the distribution of soil moisture in time, place and depth, is often a critical factor in plant growth; it can also exercise a limiting influence on activities within the soil, such as those concerning bacteria, diseases, pests and parasites. Furthermore, water erosion of soil is a problem of great importance.

The biological problems fall under three main headings; firstly the internal water status of the plant, secondly the pathways and controls affecting the transfer of water within the plant, and thirdly the response of the plant to water stress at any point in its system in respect of physiology and morphogenesis. A further category can also be included, namely the response of disease and pests to surface wetness or dryness, which can affect both the sporulation or dissemination and also control the infection conditions.

This last named type of problem is also important in regard to animal diseases, and the influence of moisture or humidity conditions on the spread of epidemics can be of the greatest importance; the control brought about by soil-moisture conditions on the life cycles of animal parasites such as flukes, ticks and nematodes has already been mentioned. A further zoological factor deals with the water-balance factors of animals. The internal water-balance is relatively unaffected by external conditions provided that adequate drinking water is available. Under circumstances of stress, however, many complicated problems arise which need careful investigation. An excess of water affecting the external surfaces of animals can also lead to loss of condition and be a contributory factor in subsequent decrease in performance or disease incidence. This "exposure" effect also involves problems of temperature and wind regimes.

Momentum

Air movement, from the wind blowing above a crop to the eddy movements within the crop stand is of the greatest importance in agro-meteorological and agro-climatological studies. This movement changes in strength with height above the ground surface, and the physical mechanisms of the eddy transfers of momentum determine the velocity profiles which fit a logarithmic form above the effective surface layer and something close to an expotential form within the crop. Not only is momentum transferred in this way, but also factors such as heat and substances such as water vapour, carbon dioxide and all air-borne particulate matter or pollutants.

The first requirement from the atmospheric sciences is for knowledge of the measurements of air movements in time and space, a requirement which is not easy to meet to the degree of accuracy and complexity which may be required for fundamental research. The variations from point to point and from second to second present a four-dimensional sampling problem of the greatest difficulty. The aerodynamical analysis of air movement near the ground is therefore exceedingly complex, and much depends on the identification of the skin friction caused by the soil surface and the form drag imposed by the variable geometry of the crop stand or animal shape.

In consideration of soil problems, the outstanding manifestation of air movement, kinetic energy and momentum is that of erosion. This threat to land potential can only be understood by a full knowledge of the physical interpretation of the meteorological and soil conditions. The danger and extent of loss of topsoil through wind erosion can only be assessed with meteorological assistance. A further, though less dramatic, function of air movement in respect of the soil concerns the gas interchange (water vapour and carbon dioxide) at the soil—air interface, which concerns both evaporation and plant respiration. The vertical profiles of such gases are logarithmic and aerodynamic formulae can be used to calculate their vertical transport.

There are three main types of crop response to wind or air movement. One is the crop geometry or deformation in shape caused by wind, to which reference has already been made. Another is the extremely important but relatively unknown effect of wind on stomatal control and on crop growth. The results of experiments on the effect of shelter and the apparent limitation to production observed in the windier areas of the world suggest that we still have much to learn about the true effects of air movement across the surface of a plant. Finally, there comes the question of physical damage, the wind-throw of trees or the lodging of plants. A growing plant, whatever its size, can have a natural period of vibration about its point of fixed contact with the soil. If this period coincides with a similar periodicity of increase of wind strength, a resonance effect is created and a breaking point can be reached.

Wind strength also affects the take-off and landing of pathogens and insects, besides having a secondary effect on temperatures and humidities which might determine infection conditions. The conditions which control the habitat and dispersion of insects are greatly influenced by the wind regime. With larger animals which have a greater choice of controlled and selected mobility, the wind is less critical but nevertheless plays a large part in determining that ill-defined variable "exposure".

Carbon dioxide

Carbon dioxide is essential for plant growth and the full examination of the processes such as photosynthesis demands measurements of the CO_2 flux and understanding of the physical processes involved. The vertical flux can be measured by placing an enclosure over the vegetation and precise results can be obtained by this method, although it has the disadvantage of possible interference with the exact environment, as the presence of the enclosure may modify the processes.

Methods based on micro-meteorological techniques are less precise, although no disturbance of the environment is involved. The difficulties lie in the exact assessment of the vertical profiles and simultaneous estimates of vertical diffusivity. Mathematical models can be devised to calculate the CO_2 exchange rates in simplified situations, but the greatest source of possible error lies in the lack of full understanding of the problems of turbulent flow of air.

Besides the measurement of fluxes and profiles, the meteorologist is concerned with the distribution of CO_2 in time and space on a global scale, Although an increase of the concentration in the atmosphere could lead to improved rates of photosynthesis, the effect of such an increase on the heat balance, the general circulation, and hence on world climate, is by no means clear. It is possible to argue that the increase would cause the general air temperature to rise appreciably, but conversely, a subsequent increase in cloudiness might have a reverse effect. Much more accurate sampling, much more reliable scientific logical thought is needed before the net effect can be foreseen. This is an urgent and important area of research which is of importance to a much wider field than agro-meteorology alone.

The soil scientist is concerned with the flux of CO_2 from the soil into the air, spatially, diurnally and seasonally. The factors which affect the magnitude of this flux are fairly well appreciated, but as yet we have no reliable means of quantifying them and so affording any means of prediction. The supply of CO_2 from the soil is not a major factor in the provision of material for the process of photosynthesis, as the atmosphere is capable of providing most of the crop needs by turbulent mixing. Nevertheless, most of the CO_2 fixed in this process ultimately returns by respiration from the soil, so that knowledge of this factor is needed in any examination of the CO_2 cycle.

The biologist has two main problems to consider; the first deals with the transport within the transpiration stream and the role played by stomates and controls of stomatal opening which aid or impede the CO_2 oxygen exchange, the second deals with the plant response to the gas in regard to both photosynthesis and respiration. To attain a sense of proportion, it should be realised that the CO_2 annually exchanged by vegetation on land is about one tenth of that exchanged between the sea and the atmosphere; as far as carbon storage is concerned, plants retain about one-fifth of the amount stored in the atmosphere, and much less than that stored in an inorganic form in the oceans.

Transfer of other air-borne materials

The list of substances, living or dead, visible or invisible, solid or gaseous, carried from one place to another by air movement is a very long one. With very few exceptions, of which pollen is one and sulphur in some areas is another, all such forms of air-borne material are liable to have adverse effects on food production, both animal and vegetable. The main sources of such noxious pollutants are the urban areas or mechanical transport, but it must not be forgotten that the land and vegetative covering thereof also provides a multitude of sources of pathogens and farm chemicals.

Conditions leading up to take-off and infection on landing for pathogens and pests have been referred to earlier. Otherwise a great deal of similarity exists in regard to transport, dispersion and deposition for all types of pollutant, the main discriminating factor for particulate matter being the size of the particle. The meteorologist therefore has a three-fold problem; firstly the distribution in time and space of the air-borne material, in other words the air quality; secondly, the transfer processes from release to deposition; thirdly, the controlling ambient conditions which determine whether a pollutant will remain viable in transit or change its nature.

The soil scientist is concerned about the distribution of sources (or originating areas) and sinks (or reception areas). The soil is the main source for subsequent transfer of pesticide residues; although this section deals principally with air-borne transfer, many such residues can travel by water, especially where heavy leaching occurs.

The biologist is concerned with release and post-deposition absorption processes; he also has to deal with the response of the crop to such absorption, and he has to be aware of the response of biological pollutants (not necessarily pollution materials) to ambient environmental conditions. Animal scientists face a similar treble list of associated problems.

The present lack of knowledge

The present status of process knowledge among the several areas described in the preceding paragraphs is very variable; there is also a lack of basic background data regarding certain aspects of meteorology. In particular, there is an urgent need to know more about the content of the air, and to a lesser extent of the water, which affects processes in the biosphere. This means to say that the distribution in time and space of such things as carbon dioxide and other pollutant gases, particulate matter such as all forms of pathogens, and biological matter such as pollen is relativley unknown as compared to the extensive and fairly accurate data concerning the physical properties of the air. Very little detailed continuous observational work has been done regarding surface conditions, especially in regard to temperature and moisture, and even within the soil itself, the moisture conditions are often a matter more of conjecture than rigid comparable measurement.

There are several other areas of relative ignorance, which are italicized in the following summary.

Summary of basic processes and data

Process subject:	Atmospheric sciences:	Soil sciences:
Radiation	Distribution of radiation *Temperatures of surfaces*	Optical properties of the soil
Heat	Convectional and turbulent transfer distribution and associated processes	Temperature distribution Conduction of heat within the soil
Water	Distribution of water input, (precipitation climate) Distribution of water output, (evaporation climate) *Wetness of surface* *Processes of dew formation*	*Infiltration* of water Internal percolation, drainage and *redistribution* of soil moisture *Distribution of soil moisture*, both in quantity and quality
Momentum	Form surface drag and skin friction Distribution of air movements near the earth's surface	Erosion processes Exchange of gases between soil and air
Carbon dioxide	Vertical fluxes and profiles within the biosphere *Distribution in time and space*	Flux from the soil
Other materials	*Distribution of air-borne matter in time and space* Transfer and dispersion factors; physical conditions during transfer	Distribution of pollutant sources and sinks

Summary of basic processes and data (continued)

Process subject:	*Plant sciences:*	*Animal sciences:*
Radiation	Plant response Optical properties and shape of vegetation *Surface temperatures*	Animal response (growth, production and mobil- ity) Optical properties and shape Surface temperatures
Heat	*Plant response* Pathogen response	Animal response (Mobil- ity, production and reproduction) Pest and parasite response
Water	*Distribution of moisture within a plant Movement of moisture within a plant Plant response* Pathogen response	Animal water balance (internal and external) Pest and parasite response
Momentum	Plant response (growth) Deformation of shape Physical damage	Animal response (mobil- ity) Exposure effects on ener- gy, balance and per- formance
Carbon dioxide	Plant response (*photo- synthesis* and respi- ration) Internal transfer	Response of soil fauna
Other materials	Plant response *Release and absorption* Pathogen response to environment	Animal response *Release and absorption* Pest and parasite response environment

REFERENCE

Smith, L.P. (Editor), 1972. The application of micrometeorology to agricultural problems. *W.M.O. Geneva, Tech. Note*, No. 119, 74 pp. (see Appendix I).

CHAPTER 2

The Materials of Agricultural Meteorology

THE MATERIAL — METEOROLOGICAL OBSERVATIONS

Meteorologists do not always realize how lucky they are in regard to the immense amount of past data available for their use. With the possible exception of astronomy no other science has a comparable treasury of information at its disposal. The reason for this is partly the fascination of the subject, partly the relative ease with which the simpler elements of weather can be observed, but mainly because of the demand for information from maritime and aviation interests. Starting a series of systematic homogeneous records is always something of an act of faith, hoping that ultimately they will be of use, almost like planting trees for the benefit of succeeding generations. Ships and aircraft, however, needed instant information and were in a position to pay for the inevitable costs; lives of passengers and crews were at stake, the need was obvious and action was taken.

The meteorologists, on their part, carried out their duties with a deep sense of responsibility. They established international standards of measurement; they checked their records with care; they published much of the data and stored it for future perusal and use. Looking back at the resolutions and recommendations of the old International Meteorological Organization, the forerunner of the present World Meteorological Organization, the Technical Agency of the United Nations, it is clear that with very few exceptions, the decisions they took regarding the acquisition of meteorological data were of the highest order and have stood the test of time. A great debt of gratitude is owed to those scientists of the 19th century who laid such a solid basis of what is now called the "monitoring" of the physical environment.

It would not be fair to say that the system was, or is, perfect. Despite all attempts at checking and counter-checking, errors creep in; the old saying that "Error is a hardy plant, it flourishes on any soil" is only too true. Standards vary, recording sites change in character, instruments can deteriorate in performance or they may be replaced by newer types with different performance characteristics. Some meteorological elements are difficult to measure, so that there are gaps in the spectrum, but the biggest gaps are in the spatial coverage, or network.

The smaller well-populated countries are the most fortunate in this respect, and their network density is usually far greater than that in other areas. Even so their reporting stations tend to be concentrated on airfields or on the coasts — a legacy of the paramount interests of shipping and aircraft. Observations and research stations generally maintain good records, but their numbers are not large and the bulk of climatological information often comes from a haphazard scatter of co-operating units rather than a planned network. This is not entirely a disadvantage, because quality is far more important than quantity; bad unreliable observations are far, far worse than no observations at all.

Before considering the individual elements, it is helpful to review the general types of reporting station. Firstly comes the type usually termed the "synoptic station", because their observations form the basis of the synoptic maps used in weather forecasting. Such stations record and report a wide variety of meteorological information at least every 3 hours every day of the year, and often every hour. They maintain a variety of autographic records and from them it is possible to obtain a very full picture of the weather on a relatively short time scale. They are, of course, the least numerous, but their standards of observation are high, they are manned by full-time professionals, and their records can be of the utmost value for agro-meteorological and agro-climatological work.

The next class of weather reporting station can be called the "climatological station", and usually takes only one set of observations each day, although some may report more frequently and may keep some autographic records. Although simpler in extent, their records can be regarded as the backbone of any regular systematic assessment of weather and climate. They are roughly ten times more numerous than the synoptic stations and many have been faithfully maintained over long periods of years.

An even higher order of magnitude is encountered in regard to stations which simply record the rainfall or snowfall, generally on a daily basis, but sometimes in remote areas, on a monthly basis. The cynic might say that the one thing a meteorologist does know is how much rain has fallen, even if he is not quite sure how much is likely to fall in the future. Although the rainfall network is the densest of all it still may fall short of investigational requirements.

Accepting that the standard of observation at official and officially supervised meteorological reporting stations is in most cases reasonably high and is generally available, it remains to examine how representative such observations can be of conditions in surrounding areas and hence what use can be made of them. Each element will therefore be considered in turn.

Pressure

Pressure is measured frequently and accurately at all synoptic stations. To say that for agro-meteorological purposes "pressure is pointless", is perhaps to dismiss its importance too abruptly. There are a few biological processes which react to pressure or more possibly to a change in pressure, especially in the field of animal, bird or insect behaviour. Certain work in the laboratory needs pressure measurements for strict accuracy, but for the most part, existing pressure observations at nearby official stations meet the requirements for most investigational or research purposes. Certainly the values and variations in pressure which are to be found in official records can be used as reference values without any great loss of accuracy. Care may have to be taken with respect to any change in height above sea level, and obviously the chosen pressure reference point should not be a great distance away. Even so, as broad influences are probably sought for rather than immediate local effects, the existing pressure data is likely to meet most requirements.

Air temperature

A very different attitude must be taken in regard to air temperature. The official reports of air temperature are not really the temperature of the air. They are the temperature of the sensing element of the thermometer which is in a white louvred box (the screen) some distance above ground level. This seeming artificiality is essential for standardization and homogeneity, but the true air temperature varies appreciably with time, with height, with type of ground surface, with the nature of surrounding objects or surrounding terrain, and a host of other factors. Air temperature, however well measured, is not the temperature of a leaf or animal surface.

Nevertheless, with all such misgivings and reservations, the value and importance of standard measurements of air temperature should not be underrated. They can serve as excellent reference values and, in particular, the deviations from average over a period of time can be applied over quite a large area. In other words, if an official station has a mean temperature two degrees above average, a neighbouring site would also be likely to be about two degrees above its average, whatever that may be.

The most rapid changes in air temperature (at standard height) occur with change in height above sea level. Such changes vary according to whether the change is upwind or downwind, but in general mean temperature decreases with height. Minimum temperatures, on the other hand, may increase on certain nights, when descending into a valley. It is thus possible to extrapolate a regime of mean temperature (maximum plus minimum divided by 2) with a greater degree of confidence than when dealing with extremes of maximum or minimum.

Other significant changes occur in the immediate vicinity of large areas of water such as the sea itself, estuaries and lakes. These changes may be very significant in regard to biological processes, and the general use of standard observations should be confined to inland situations, with as little change in land nature or topography as possible.

The standard height and customary exposure of official instruments varies slightly from country to country in spite of international regulations. The differences are small in effect but their existence should be recognized. Errors in measurement are infrequent but 5-degree (or even 10-degree) errors in maximum values are by no means uncommon. Most systems of checking eliminate obvious errors such as these, but it should not be assumed that all records are immaculate.

Provided that a cautious and intelligent attitude is maintained, most air-temperature records from synoptic and climatological official stations can be used with a considerable degree of confidence and much use can be made of them. Certainly, agro-meteorology would be lost without them.

On the other hand, autographic records such as thermograms should be used with a great deal of care, particularly in respect of extreme values. They can also give misleading values when it is required to know the time above or below a given temperature threshold. Whenever possible, hourly values should be used which have been taken by an observer. Accurate recording needs more expensive and sophisticated equipment than is usually available at a normal observing station.

Surface temperatures

To put it politely, the standard of air-temperature measurement on or near the surface (usually minimum temperatures) falls far below that of other meteorological elements. Usually this measurement is referred to as the "grass minimum" and is defined as the temperature recorded by an alcohol minimum thermometer, unscreened, laid horizontally with its bulb touching the leaves of the grass. To maintain such an exposure, the thermometer would have to be moved up each day as the grass grew and down again when it was cut, which of course never is done in practice. The definition also conveniently ignores the fact that grass, especially green grass, is not the normal ground coverage of the vast majority of the countries in the world.

In practice the thermometer is probably mounted on short supports at an unspecified height, and often not over grass at all. In some cases it is laid on the surface, the nature of which changes with the seasons. In short, a degree of untypical chaos exists, and any readings of this nature should be treated with extreme caution. Some use can be made of them to indicate the nights on which there was a large difference in minimum temperature between screen height and earth surface, but as four such grass minimum thermometers exposed in close proximity to each other will in all probability give four different readings, the precise numerical values of such measurements are hardly dependable.

It is always difficult, if not well nigh impossible, to alter an established practice, even if the practice is a bad one. Nevertheless, it is possible that some form of standard surface minimum reading could be devised which would be definable and representative. An alcohol minimum thermometer laid flat in contact with a standard concrete surface does give reproducible results. The reading therefrom would be no more artificial than an air temperature read within the usual wooden box, and at least the exposure would be similar at all sites. Any such form of desirable standardization lies in the future, and at present it would be wise to discard any records of surface minima or number of ground frosts, or at the very least, to treat them with extreme caution as a guide to circumstances rather than a scientific specification.

In practically no case is it safe to extrapolate surface temperatures.

Air humidity

The humidity of the air is best measured by taking simultaneous readings of dry bulb and wet bulb thermometers and then using hygrometric tables to give either the relative or absolute humidity. The usual hygrograph using human hair or similar material as a sensing element is subject to a variety of errors and unless very carefully maintained and constantly calibrated, can give misleading results.

Absolute humidity is fairly conservative over flat land areas above crop level but the relative humidity is sensitive to changes in air temperature, so that again it is difficult to extrapolate humidities determined from screen-readings, especially within a crop or a tree plantation, and inside buildings or shelters. Nevertheless, hourly humidities taken at synoptic stations can be of considerable use as reference data. Humidities taken once a day at climatological stations have very limited potential.

Wind

Wind force and direction vary rapidly with time and height, and appreciably with site and exposure. It is therefore very difficult to measure in a standard uniform manner. To eliminate indefinable effects and to provide a representative sample, official stations expose their anemometers some considerable height above the ground (usually 10 m). Again this provides useful reference data, but in no way can be considered to be the wind experienced by a plant or animal.

If it is only required to seek evidence of strong winds, such records may be taken to represent conditions over quite a large area, as gales rarely coexist within close proximity of calm weather. Moreover, if a relative picture of the wind regime for an area over a period of time is required, the official data may be adequate for the purpose. In particular, the autographic records (or the hourly reports) of wind direction and force can sometimes be of the greatest assistance.

Some climatological stations use cup-counter anemometers which record run-of-wind and expose them at lower heights. Theoretically these may be of more immediate use than the more detailed anemograph records, but care must be taken regarding upkeep of the instrument and the nature of its exposure. The readings are of far more local significance and no information is provided concerning wind direction.

Standard wind data are therfore of limited use, but can be valuable at times. The conditions under which the observations are made and the degree to which they can be used away frjm their actual point of observation must always be carefully considered.

Precipitation

As has already been pointed out, the meteorological element which is most widely measured is precipitation, usually in the form of rainfall. The rain-gauge used for this purpose can vary in size and in height of rim above the ground surface, both factors affecting its efficiency of catch and hence influencing its value as an aereal sample. As a result it is doubtful whether rainfall measurements are quite as accurate as some people imagine, even for the point of observation, and a radius of error of some 5% is not impossible. The effect of wind regime above the gauge can be of great importance, and particular care has to be given to the question of exposure in upland areas.

Long-term values of rainfall totals, or averages, can be often quite representative of a uniform local area, but totals of single storms or short periods such as a day can only with safety be regarded as referring to the point of observation. Details of rain duration and rainfall intensity are also only strictly applicable to their point of origin. Useful guidance can be obtained from such records, but they cannot be expected to represent the conditions elsewhere to a high degree of precision.

The measurement of snowfall is a more difficult problem and the degree of inaccuracy is greater, especially in respect of snow in mountain areas. As about 10 cm of snow is equivalent to about 10 mm of rain, the actual errors are less than the apparent errors, but problems of snow drifting and efficient catch by the gauges are by no means completely solved.

Rainfall varies appreciably with height and stations comparatively close to each other, but differing in height above sea level, can have very differing regimes. Noticeable variations over flat country can also be recognized in the vicinity of a sea coast, but these differences are of a lesser order of magnitude.

Rainfall data, because of its multiplicity, has many uses in agro-meteorology and especially in agro-climatology, but individual measurements should always be made in connection with specific experiments.

Sunshine

Most synoptic and climatological stations measure incoming solar radiation in terms of hours of bright sunshine. This is a type of "off—on" measurement which records nothing at all if the sun is clouded over or weak in intensity, and gives a time measurement when such intensity exceeds a modest limit. Errors can occur in the interpretation of the record, usually a burn on a specially prepared card, which means that the time of day as well as the duration of the sunshine can be ascertained.

Despite its obvious limitations, a great deal of information of practical use can be obtained from such records. They are fairly representative of a considerable area around the point of observation, except near a sea coast, where relatively large variations can occur over small distances. Sunshine also decreases with height above sea level, but areas in the lee of hill or mountain ranges can often experience an increase in sunshine hours, in much the same way as they have lower relative humidities.

Additional weather information

In the absence of sunshine recorders, observations of cloud amount taken at frequent intervals can be made use of in some agro-meteorological work. The type of cloud reported can also be helpful in indicating the presence or absence of convection or inversion levels. In the same manner, visibility records which are primarily required for aviation purposes, can give some indication of conditions which are of significance to agricultural problems. The presence or absence of fog, for example, may be very important in regard to disease incidence. Similarly, notes made by official observers concerning dewfall can be of great value.

Extreme weather such as thunderstorms, hailstorms or short periods of heavy rain are reported by synoptic stations as they occur and can also sometimes be traced in climatological station records. Because of their local nature, they have limited use but their occurrence does indicate the existence of a certain type of weather conditions, knowledge of which can be useful.

Details of frost occurrence, intensity and duration can be obtained from such stations, but these cannot be extrapolated with any confidence except in very flat terrain: once again the official station gives a general guide rather than precise information.

Weather details from climatological stations which take observations only once a day are of more limited value. Much will depend on the enthusiasm of the local observer, his type of employment and time available to record much more than the standard requirements.

Special observations

Certain stations, both synoptic and climatological, take observations of especial value to agro-meteorology, either because of the direct interest of the station itself (as at a research station) or because they have been incorporated into a network design. Outstanding of these are measurements of soil temperatures, at various depths, and under grass or bare soil. Few stations, however, measure the daily maximum and minimum soil temperatures, which are most important at the shallower depths.

Standard soil-temperatures are very useful as reference values, and deviations from average are conservative over quite large areas. Absolute values will vary with soil type and with crop cover or cultivations, but these drawbacks are not so serious as would at first appear. The more significant changes come from variations in aspect and exposure, and with changes in height above sea level.

Soil temperatures at (say) 30 cm depth have an additional advantage in that they tend to eliminate the short-term variations of air temperature and, so to speak, integrate the current thermal regime.

Soil moisture, on the other hand, is rarely measured as a regular procedure, and even when it is attempted, the results can be unreliable and certainly cannot be regarded as representative of anything other than the point of measurement. It is not easy to carry out such measurements, and even if use is made of modern methods (such as nuclear scattering) it is doubtful whether any network of observations would attain adequate sampling efficiency.

The most encouraging progress in recent decades has been in the increased number of stations which take reliable measurements of solar radiation. Although difficulties still exist regarding the autographic recording of spot values, so that a degree of inaccuracy of up to 10% must be expected, the general improvement in knowledge regarding the radiation balance has offered a new source of data of the greatest importance in agro-meteorological research and investigation. Care has to be taken regarding instrument maintenance, including regular calibration, for if the quality of measurement deteriorates, the result is worse than before, when only estimates were available.

As regards representativeness, the remarks made concerning sunshine records apply equally to radiation measurements, but it must be stressed that these measurements must be as accurate as possible. Good radiation records easily surpass sunshine data, but bad information of this nature is far more misleading than the simpler records.

A limited number of special observations are concerned with the quality of the air, either by direct air sampling for pollutants or by chemical analysis of rainfall. Such observations are few in number at present, but may increase in the near future with the increased international concern with environmental quality. Whatever arrangements have been made or are to be made in this respect, it is essential that accuracy of measurement takes priority over number of observing points. The fewer the number of reporting stations, then the more important it is that quality is maintained. If observations increase, they must not increase at the expense of such quality.

Finally, mention must be made of observations which are intended to measure evaporation and transpiration. It cannot be emphasized too strongly that any instrument designed to make such measurements can do no more than show or record its reaction to the environment within the limitations of its design and maintenance and under its particular form of exposure. This is essentially true of all instruments, but it is very important in this context. Measurements of evaporation, however made, may be used as a reference; they can never be used as a standard against which other instruments or methods of estimation can be checked. Comparison is possible, verification is impossible.

It is probably true to say that the degree of satisfaction regarding any meteorological measurement is inversely proportional to the number of methods or types of instrument available. If one type of instrument is internationally used then it probably works very well; if many types exist then it is surprising if any one of them is good. This latter state of affairs is the case with evaporation measurements.

The existing types vary from instruments such as lysimeters buried in the ground and filled with soil, to tanks of water also buried in the soil, to pans usually of smaller size which hold water and either rest on the soil or are supported at some height above ground level. There are also types known as atmometers and evaporimeters such as the Piché, which is essentially an inverted test-tube and a small evaporating sensor at the base. All suffer from one very difficult problem, that of defining the exact type of exposure which will ensure a homogeneous set of data at any one place, and comparable data in a network of sites.

The reason for this is that the water-vapour flux upward from any surface depends on the profiles of temperature, humidity and momentum in the immediate vicinity of that surface. To attain any stability in such profiles the upwind fetch of the airflow must be over fairly uniform terrain. This requirement implies the use of "guard-rings", so that a crop lysimeter gives the most representative readings in the centre of a large field, and an evaporating pan or tank works best in the centre of a lake or reservoir, a state of affairs rarely possible in observing practice.

Any form of tank or pan is an attraction to birds and animals as a source of drinking water; any form of wire protection to exclude such degradations alters the measurements (recent experiences in Israel suggest that such wire covers reduce the evaporation by about 9%). The list of other difficulties, such as hidden leakages, is almost endless, but most of all comes the potential error when the measurement depends on subtracting one large figure from another large figure. The difference between (say) 20.5 and 20.0 is unquestionably 0.5, but if 20.5 and 20.0 are subject to 1% errors the subtraction could be $20.7 - 19.8 = 0.9$, or $20.3 - 20.2 = 0.1$, an effective error either way of 80%.

In effect, therefore, the measurement of either evaporation or transpiration is a very difficult process. Any data obtained in this way is representative only of the instrument used and the site in question; extrapolation is a very hazardous procedure.

Meteorological working material

Much of the working material of the weather forecaster can be of help to the agricultural meteorologist, in addition to the basic data. Chief among such material is the synoptic map, drawn up at regular intervals both at surface and upper-air levels. Apart from giving the synoptic picture of the weather and the changes thereof, the wind patterns which are inherent in such maps are the basis of the air trajectories which are so important in many applications of aerobiology. The essential information is there for the using.

In addition to knowledge of winds at various levels, the radio-sondes which take vertical samples of air temperature and humidity can be of considerable use in specific investigations of an aerobiological nature. It is probably true to say that more data is available regarding the conditions of the air above the earth, than is available for conditions within the ecosphere, that is to say, the infinite local variations within crop stands.

The main purpose of such information is, of course, the preparation of weather forecasts, but that is no reason why it should not be put to further use after its immediate digestion by the forecasting machines. The weather forecasts themselves cannot be regarded as data in the strict sense of the word.

Summary

A brief summary of standard data and its usefulness can be made as follows:

	Generally useful	Limited use	Treat with caution
Generally available	Air temperature Precipitation Sunshine Synoptic maps	Pressure Wind General weath- er data	Surface temperature Air humidity
Limited availability	Soil temperature Radiation	Air quality	Soil moisture Evaporation Transpiration

THE MATERIAL — DERIVED DATA

Derived data, in this context, implies the type of information for a site which is not measured at that site, but which can be deduced to a reasonable degree of accuracy from such observations that are made on site. It does not refer to extrapolated data which implies that observations at one place are used to estimate the conditions elsewhere. Three types of derivation can be envisaged; one type deals with a single meteorological parameter, another deals with a combination of parameters, and the third is concerned with the estimation of a parameter not basically measured by the use of calculations involving parameters which are measured.

Single parameters

Air temperature, soil temperature, humidity and wind are measured at specific times under standard conditions of height and exposure. It is often necessary to make use of values of such parameters at different heights or at different times. If autographic records are kept, interpolation in time (without variation in height) presents no difficulty. The estimation of values at other than the standard heights may be much more difficult, especially in the vicinity of the ground. Such estimates may have a relatively large radius of error, but nevertheless the values so obtained could be of an acceptable degree of accuracy for the problem under consideration.

Mean values of observed parameters are not always available, and if they are part of published data, the time-period of the averaging process may not be suitable. For example, the mean air temperature over daylight hours is not normally available, but can be derived from existing data, especially if hourly values are known. Monthly means are commonly used in climatological statistics; the plant, however, has its own calendar and reference may have to be made to daily values to find the information over a significant period. Such requirements may be time-consuming, but they rarely present any theoretical difficulty.

A similar state of affairs exists when it is required to know the duration of time when a given parameter threshold is exceeded or not. If autographic continual records are kept, and are accurate, then there is no difficulty; if they are not kept then a degree of interpolation has to be attempted. A common example of this type is the use of accumulated temperatures, which are the temperature—time base integral above or below selected thresholds. Such finite integrals are sometimes referred to erroneously as "heat sums", which they most definitely are not.

Much derived data of single parameters can therefore be obtained by the simple arithmetical processing of existing material. On other occasions, however, a special relationship between the actual data and the required data must be established, either by recourse to theory or by setting up a pilot study of the required comparison. Such a pilot study may use the experience of another station where both sets of data are measured, or it can involve the setting up of a short-term experiment to clarify the relationship at the site in question.

Combined parameters

Examples of this type of working material are: frequency of strong winds from a given direction; of high/low temperatures with high/low humidities; of frosts with winds above a given strength; of rain with wind and/or low temperatures; of wind direction and stable or inversion conditions in the lower air layers.

The type of combined parameter varies with the nature of the problem which is being investigated. In all cases, it is unlikely that the information will be available in a recorded or published form, unless it has been extracted for a similar purpose on an earlier occasion. It does therefore demand a degree of extra work, but such work is likely to be simple in nature. The main disadvantage is the time needed to do such work, which may be considerable unless an efficient modern system of data processing is available.

No expensive or time-consuming multiple analysis of this type should be undertaken until it is fairly certain that the chosen combined parameter is a useful one. This is not always easy in the early trial stages when the search for the parameter is going on. However, once a combined parameter has been established as a helpful summary of significant conditions, then machine handling of data will save time and effort.

Calculated parameters

If certain aspects of weather, or the consequences thereof, are not observed or measured, or even if the measurements are subject to a considerable degree of justified doubt, it may be necessary to estimate the parameters by some form of calculation. Examples in this category are radiation and light, evaporation and transpiration, soil moisture and surface wetness.

Estimates of incoming radiation or light can be made from observations of sunshine hours or even from cloud amounts, and many formulae have been published which aim at providing such estimates. Latitude and season of the year have obviously to be incorporated into such formulae which sometimes also include other variables such as atmospheric turbidity (the dust or particle content) or visibility. Errors are bound to be introduced in such a process, arising either from the nature of the basic information or from the inadequacies of the conversion formulae. Nevertheless, the results so obtained may be of sufficient accuracy for the purposes of the investigation. The possible dangers and misleading effects of errors are minimized if the radius of error is understood and allowed for in subsequent use.

A similar state of affairs, a multiplicity of formulae, exists in regard to calculations of evaporation or transpiration from observed meteorological data. They range from the simple to the complicated, from almost straight empiricism to a near approach to scientific calculation. Most formulae of this nature are fairly reliable under the circumstances which gave rise to their birth, although their validity does depend to no little extent on the ability of their originator, or sometimes on his liability to self-deception. Generally speaking, the designer of the formula has a clear idea of its strength and limitations, but the same cannot always be said of his disciples. Techniques for calculating evaporation have been likened to religions, in that the basic idea was eminently suited for the environment under consideration, but that subsequent developments and over-enthusiastic users can lead to a state of affairs which can only be described as "the best of intentions, yet the worst of results"; the soundness of the basis has been lost in a wealth of irrational detail.

Soil moisture estimates depend a great deal on the accuracy of such evaporation or transpiration formulae, but also on a number of other rather imponderable factors. The formulae generally estimate potential transpiration, and it is the actual moisture loss from soil and plant cover that is the desired parameter. The conversion from potential to actual is probably the biggest source of error, but other factors such as soil type, crop type, and questions of aspect and exposure can be of major local importance. Under these circumstances, precise accuracy cannot be hoped for, but a degree of acceptable accuracy, again with an awareness of the radius of error, can be attained by the careful worker.

Surface wetness is really a special case of soil-moisture estimation. Reasonable estimates are generally easier to obtain, especially if the rainfall data is adequate in regard to the time factor. In other words, 24-h totals of rainfall are not good enough for this purpose, although 12-h totals, separating roughly the day rain form the night rain can often be sufficient.

It has been found that the duration of wetness of leaves after rain can be estimated with a fair degree of precision by counting the number of hours the humidity (in a standard screen) remains above 90%. Results obtained in this way compare very favourably with measured values.

The details of the required calculations for the various parameters will be discussed on p. 45 ff., but a few general observations are appropriate at this stage. As has already been stated, any method can be regarded as helpful, within its inherent limitations, under the circumstances in which it was formulated. It is when such formulae are used uncritically in different climates or localities, that difficulties may arise and results become meaningless. Empirical or even semi-empirical methods are unsuitable subjects for export; home use may be successful, use elsewhere cannot guarantee results unless some form of verification has justified their extended validity.

Moreover the formulae themselves must not be misused, even in their area of origin. If they were designed for a given time-interval, say a month, they cannot be used with daily data without running the risk of major error. For example, a method of calculating the mean monthly radiation from sunshine data may produce worthwhile results; it might even give acceptable answers for 10- or 15-day periods, but it would be very unreasonable to apply it on a daily basis. Similarly for evaporation formulae, which are often based on monthly data, daily values cannot be depended upon without additional evidence.

Before any formula which has been successful elsewhere can be introduced into a different environment, such verifying evidence must somehow be obtained. This can be extremely difficult, especially when verification against imperfect observations is a doubtful procedure, or when no observations exist. It is a vicious circle in which you should not use a formula without verification and yet cannot verify it without using it. The only answer is to try to obtain verification by deduction, so that the formula is accepted as a hypothesis and then tested by comparing the theoretical consequences of application of the formula with field evidence. In other words, does the theory fit the facts? If the calculated values are found to provide a practical means of explaining a circumstance, then they are manifestly useful in that context, even if the absolute values of the derived parameter are still short of absolute accuracy, and still unproven.

ESTIMATED DATA

If no meteorological observations have been taken at a given site, but are available from a standard reporting station in the neighbourhood, it may be necessary to use some form of extrapolation, translating the observed data into assumed data at the place in question. This can only be avoided if the standard-site observations are merely used as reference data and it is assumed that their relative variations are typical of the site where no observations are available. Clearly in this case, no reliance can be placed on absolute values.

The process of translation is made much easier if there has been, at any time, a series of parallel observations at the regular site and the temporary site. From side-by-side analysis of the two parallel sets of data, conversion factors can be devised which form the basis of extrapolation. The complexity of such conversion factors will depend on the degree of comparison which is possible and the requirements for data at the temporary site. The methods employed will vary with the nature of the parameter.

If no such parallel records exist, a greater degree of uncertainty will be introduced in any extrapolation, but even so it is possible to make reasonable estimates, based on theoretical grounds, for some meteorological variables. As a general rule the reliability of such estimates is far greater in regard to long-term means or averages. Estimates of climate are thus often fully acceptable as a basis for investigation. It is the short-term values of a parameter, the day-to-day weather, which is most difficult to estimate with any degree of certainty, although again this varies from parameter to parameter.

If more than one standard station with a regular observing sequence is available for reference, the process of extrapolation becomes one of interpolation or of double (or treble) extrapolation, thus providing a better hope of accurate estimates. Any such process should involve a knowledge of the causes of variations in meteorological values; a simple linear interpolation is rarely possible using direct observed data. If such data have been reduced to a common datum line, as for example a reduction to mean sea level, which has removed the dominant site-to-site difference, then linear interpolation becomes a more practical method of estimation.

The most easily recognized factor which causes a change in the value of a meteorological variable from one location to another is height above sea level, but proximity to the sea or any other large expanse of water can cause rapid changes with distance, or in other words a gradient. The smaller the gradient, the easier it is to make an estimate; with strong gradients, the probability of error is far greater.

Pressure

Average pressure, if ever it is required, is obtainable with adequate accuracy from maps of such averages reduced to a standard level of height. Daily values, however, present more difficulty unless daily pressure maps are available. If they are not, and only one local station is available, account must be taken of wind speed and direction to estimate the pressure gradient. In most cases, however, absolute values are not required and it is sufficient to use the local standard station as a reference point.

Air temperature

Mean daily air temperature is far easier to estimate than maxima or minima. It is, in fact, very unlikely that adequate extrapolations of extreme temperatures can be made unless the sites are close together on level terrain, or unless a series of parallel observations are available for comparison. Minimum temperature in particular is a very unsuitable parameter to treat in this way, and the greatest care must be taken to establish the validity of any method of estimation.

The average variation with height above sea level of mean air temperature in any region is usually known approximately. It does, however, vary with season and with type of weather, so that in some circumstances the generally applied correction may not be adequate. Depending on the degree of accuracy aimed at, it is always worthwhile to examine the nature of the height correction and determine whether the possible deviations from the standard correction are of possible significance within the context of the exercise.

The process of extending the value of a short series of observations by comparison with a longer series at one or more other places is known as "weighting" and is a common practice in climatology. As far as temperature is concerned, the working parameter is usually the difference in temperature between two places; the use of ratios is not possible because of the arbitrary nature of the years on the temperature scale.

In general it may be assumed that mean air temperatures, and especially their time-based averages can be estimated with a fair success, unless circumstances demand a high degree of accuracy.

Soil temperature

It is far more difficult to extrapolate observations of soil temperature, except at considerable depths below the surface. The reason for this is that so much depends on the nature and condition of the soil, the nature of the surface cover, and the aspect and exposure of the site. The height correction is also less well established than for air temperature, and therefore it is wise to use nearby standard observations as reference values and not as a basis for extrapolation.

Soil moisture

It is well-nigh impossible to extrapolate any measurements of soil moisture, and any attempt to interpolate within a network is fraught with difficulty. Used as reference values they may be found helpful, but there is an inevitable radius of error and the effective "unit" is more likely to be one significant figure than anything more precise.

Air humidity

Absolute humidity is fairly conservative over moderate distances, so that relative humidities can be estimated to some extent with the use of estimated temperatures and assuming a constant dewpoint. The major errors will occur on "radiation nights" in areas where fog and frost are most prevalent. Generally speaking, estimates of relative humidity are not very accurate and rarely worth the trouble involved. In particular, duration of high relative humidities can rarely be estimated except over very level terrain.

Rainfall

Rainfall averages (monthly or annual) for a place for which there are no observations can be estimated to a fair degree of accuracy with the help of expertly drawn isohyets on a suitably scaled map. The estimation of current values is a far more difficult problem; for an estimation of a monthly rainfall total it is usually assumed that in a limited area the percentage of average will be approximately constant. If this percentage is known for a nearby standard station, the average for the required site is estimated and the same percentage is used. The accuracy of such a "weighting" is improved if more than one reference station is available. The main errors occur during the times of the year when showers, as distinct from continuous rain, provide the bulk of the rainfall. Heavy showers covering small areas can completely disturb the assumed uniformity of the rainfall pattern; they may indeed evade the standard network altogether. Estimates under such circumstances can at times be very misleading.

The same disadvantages apply with even greater force when considering daily rainfall totals. In the majority of circumstances there is no alternative to local observation or recording. This is especially true when rate of rainfall is required. Knowledge of the occurrence of rainfall or the possibility of showers can be obtained by extrapolation but again a degree of uncertainty is often present.

Wind direction

Details of the general surface wind direction can generally be satisfactorily estimated by reference to a nearby station or to a synoptic map. Exceptions occur when the wind speeds are low, or when pronounced alterations in airflow are occasioned by the local topography. Areas near the coast affected by sea breezes are also zones where estimates can be in error. The depth of penetration inland of such air movement varies from day to day and little reliance can be placed on any form of extrapolation. In times of strong winds, however, the direction of flow attains a reasonable uniformity over a local area, subject, of course, to its own fluctuations within the main air stream due to turbulence.

Under near-calm conditions, local winds caused by katabatic (down-slope) or anabatic (up-slope) winds can be of considerable significance in agro-meteorological problems, especially in regard to investigations concerning the spread of pests or diseases. The part played by katabatic winds in the incidence and intensity of frost is another case where little help can be gained from a standard observing network unrelated to the actual problem.

Wind force

The strength of the wind cannot usually be estimated with any degree of precision. Given a set of parallel observations, an attempt can be made to extrapolate from a standard site for such values as run-of-wind on a daily, weekly or monthly basis, but spot values are impossible except within very broad bands of uncertainty. As in the case of rainfall, local observations are a necessity if the force of the wind is needed in an accurate form.

Sunshine and radiation

These variables, depending as they do, on the position of the sun and the cloud amount are less variable from place to place than items such as temperature or wind. Questions of aspect and local shading have, however, to be taken into account and the neighbourhood of large towns or cities can create smoke concentrations which will have pronounced local effects.

In general, however, as in the case of rainfall, averages can be estimated from standard networks and in any period such as a month or a year, the percentage of average is a useful guide for the provision of estimates. Even daily values can be estimated if a high degree of accuracy is not required. The measurements themselves are by no means precise, so that the further errors introduced by extrapolation must lead to a very approximate answer. If this degree of uncertainty is inadmissible, then local observation must be made, especially in the case of solar radiation. A 7-day period is probably the least period of time for which estimates should be used in most investigations.

Evaporation and transpiration

Measurements of evaporation and transpiration are greatly dependent on the type of instrument, and most especially, on its local exposure. Such measurements are not recommended as a basis for extrapolation, although some information can be gained from monthly or annual variations expressed as a percentage of average.

Calculations of the values of these parameters are probably more useful provided that due allowance is paid to their radius of error. From the nature of the assumptions made in the construction of the several formulae, they are more likely to be applicable to the mean of a local area rather than a spot value.

The scale of variation from place to place is relatively small so that estimates of averages can be made after due regard to dominant variables such as sunshine and height above sea level. Current monthly values can also be extrapolated, generally to a degree of accuracy acceptable for practical problems on a field scale. Daily values, however, obtained in this manner can be subject to quite large relative errors although the absolute error may be small and therefore admissible. In general it is unwise to use this approach over periods of less than 10 days.

Summary

As a rough guide to the various processes of estimation, the following summary shows the general classification of the parameter.

Note: (a) In the case of elements "normally unsuitable" it is at times possible to run a series of parallel observations from the analysis of which methods of extrapolation can be devised.

(b) Any attempt to estimate daily values is liable to error, especially in respect of rainfall during a showery type of weather.

Acceptable without modification or with a site correction	Modification by:		Normally unsuitable for extrapolation
	Duration from estimated average	Percentage of estimated average	
Pressure Duration of winds of moderate strength or above	Mean air temperature Calculated evaporation or transpiration	Rainfall Sunshine	Extremes of air temperature Soil temperature Soil moisture Humidity Wind force Measurements of evaporation or transpiration

AREAL VALUES

It is often difficult to answer the question concerning the average value of a meteorological parameter over an area. If, indeed, the parameter varies appreciably over the area concerned, it is a matter of considerable doubt as to whether the answer, whatever it is, has any valid meaning. Nevertheless, if it is wished to compare, say, the yields of a particular crop meaned over an area, with weather factors, some attempt must be made to obtain an areal figure.

It is probably true to say that the most justifiable method of finding an areal value is to plot all available observed data on a map, then with considerable care and insight to draw isopleths around such data, finally using a planimeter to find the areas between each isopleth. This being done, it only remains to find the average from the weighted mean of the values appropriate to each area. This process takes time, and is very dependent on the skill and accuracy with which the isopleths are drawn. The one important variable which can be treated in this manner is height above sea level. The contours of height on a survey map are accurately drawn, and the only sources of error thenceforward is in the use of the planimeter and the subsequent arithmetic.

A modification of this method is to superimpose a fairly close square grid over the area and from the observed data estimate the value of the parameter at each intersect on the grid. An areal value can then be found by using these intersect values or by taking the mean of the four in any one square as representative of the areal value of each square. Again much depends on the degree of skill used in estimating the values at the grid points, and clearly this process can be very difficult if there are few original observations.

A method which is not recommended is the unsophisticated use of irregular grids. One such method involves the construction of polygons formed by the intersection of the perpendicular bisectors of the lines joining observing points. The observed value is then taken to be appropriate to the area contained by the polygon in which it is situated. The only circumstance when this method can approach an acceptable answer is when there is very little variation in the values of the weather factor. If the cause of the major variation in value, for instance the height above sea level, is removed, then the use of such grids provides a convenient weighting factor which avoids the possible errors invovled in a data network with badly distributed points of observation. Even so, the regular grid interpolation method also incorporates this improvement and avoids the difficulties which arise when a large number of observations are concentrated in one part of the area concerned, with a sparse distribution elsewhere.

Meteorological parameters vary with height to a greater extent than any other physical factor, except perhaps in the immediate vicinity of the coast where distance from the sea becomes important. The advisability of a reduction to a standard height (generally sea level, but an intermediate height may be preferable) before attempting to find an average value is therefore clear. Crops, however, do not grow at a theoretical sea-level height, but at the height of the field concerned. A re-adjustment to average height is therefore necessary after determining a "sea level" areal value.

Much now depends on the accuracy of the corrections used for height. Although the conventional use of standard corrections for, say, the variation of temperature or pressure with height has much to recommend it, it is always advisable to establish a height correction value appropriate to the area and to the season, especially in respect of rainfall. This can normally be done by intelligent uses of existing long-term averages, although if they do not exist a degree of improvisation is inevitable.

Similar major adjustments, prior to averaging, can also at times be made if there is a definite gradient with distance over the area concerned, as for example a general change independent of height in any one direction. Sunshine, for example, often decreases with distance from a sea coast. The nearness to a range of hills or mountains can also have a pronounced effect on spatial variations, especially as regards the so-called "rain shadows" in the lee of a range. If these changes can be quantified, the process of estimating an areal average is made easier, although the readjustment back to the mean value of the influencing factor is always necessary.

The process can now be summed up as follows.

(a) Find the correction factors for height appropriate to the area and to the time of year concerned.

(b) Find any other valid correction factor with distance from a dominating feature.

(c) Apply these corrections to the observed values.

(d) Mean the corrected values either by a grid method, or by direct averaging if the distribution is adequate and uniform.

(e) Re-adjust the result to the average height of the area.

(f) Re-adjust again the average value of any distance factor used in (b).

The whole process is a long one, but the results may justify the time taken. It is always wise, however, to see what can be done by the use of one or two reliable sets of values as reference points. The degree of approximation inherent in a problem may not require anything further than this for practical purposes.

POTENTIAL TRANSPIRATION

This parameter is often referred to as potential evapotranspiration, thus implying a possible total loss from the earth's surface. This is a pity, because it tends to confuse the issue. For clarity of thought, it is preferable to think of evaporation from a water surface, potential and actual evaporation from a soil surface and potential and actual transpiration from a crop surface, even though at times it is convenient to combine water losses from plants and the surrounding soil.

There are a multitude of meteorological formulae for the calculation of potential transpiration, most of which have been devised by scientists to the best of their ability using such data as they had available. Most of such formulae attain a degree of accuracy in their area of origin, but most have to be used with care in different climates.

In a given climate, these formulae will give similar but significantly different results; any attempt to measure evaporation or transpiration will also give a differing set of data. What must be remembered is that no data, calculated or observed, has any claim to perfect accuracy. The different answers can be compared, but cannot be verified by reference to each other, because there is no absolute standard.

The choice of formula is therefore difficult but often there is one overriding factor, namely the extent of available meteorological data. A formula such as Penman's which requires knowledge of sunshine, temperature, humidity and wind, cannot be used unless such information is available from accurate reporting stations, preferably taking more than one observation a day; data from once-a-day climatological stations may be inadequate, radiation data will be an improvement on sunshine records. If meteorological data is incomplete a simpler formula will have to be used.

The choice of formula is in the nature of an hypothesis. If deductions are made from such a hypothesis, coupled with knowledge of rainfall, soil and plant behaviour, results will be obtained regarding certain observed agricultural or hydrological facts. If the results do not fit the facts, then either the formula or its method of application to the problem is wrong. The only true proof of the accuracy of any calculation of evaporation or transpiration is the extent to which it can be used to explain biological processes.

In the United Kingdom, it has been found by a variety of agro-meteorological and agro-climatological tests that Penman's formula gives satisfactory answers, and it is therefore used for that purpose. This does not imply that it would work equally well in other climates, or that other formulae would not give better results in such conditions. It must be used in the sense in which it was devised. It should not be used for daily values and it cannot be expected to give accurate results for small areas.

SOIL-MOISTURE BALANCES

One of the ways of using, or verifying, estimates of potential transpiration, is in the construction of soil-moisture balance sheets. This can be done on a monthly or weekly or daily basis, using appropriate data for rainfall and potential transpiration, but first a model must be devised for actual transpiration as distinct from the potential.

This model must take into account the characteristics of the crop and the soil, but the mistake must not be made of concentrating too much on valid but second-order magnitude factors which complicate a model without adding to its accuracy. The first decision to be made is the maximum amount of water that a crop is able to extract from a soil. Experience has shown that in England this is approximately 125 mm over farmland for crops such as grass. For sugarbeet or cereals, and certainly for trees, this is likely to be an under-estimate, but it appears reasonable for plants whose roots lie mainly in the top 60 cm of soil. As soil moisture is depleted and tensions increase the actual transpiration falls below the potential. An approximate model for grass can be taken as follows:
Actual equal to potential for the moisture in the top layer, say 50 mm.
Actual equal to half potential in the second layer, a further 50 mm.
Actual equal to a quarter potential in the third layer, a further 25 mm.

Other models could similarly be devised; for example, on poor land with limited depth of soil, all values could be halved giving full rate for 25 mm, half rate for 25 mm, and quarter rate for 12.5 mm. For cereals the values could be increased.

The potential transpiration calculated by the Penman or similar methods assumes full crop cover and limited vertical extent of the crop. Adjustments may thus have to be made for limited crop cover in early stages and for the height of crops at later stages. For grass, however, these modifications are not necessary and a simple procedure can be adopted.

A start has to be made at a point in time when the soil is known to be at field capacity. Thereafter, if transpiration exceeds rainfall, moisture will be extracted from the top layer of the soil until this layer is dry, subsequently extracting moisture from the second layer (at half rate) and so on. If rainfall exceeds transpiration when the soil is at capacity, the excess will be drained away. If the soil is not at capacity, the excess will be added to the top soil layer. The important principle is that moisture is always first taken out of the highest available moisture layers and is always put back into the highest available layer. Soil cannot be wetted to half capacity, the incoming rain raises the top centimetres of soil to capacity before it begins to add moisture to the layer underneath.*

It is therefore convenient to keep a balance sheet in respect of the three layers of the model. To do this on a monthly basis makes certain generalizations in regard to the distribution of the rain, but this is inevitable for a first approximation. Two examples can be quoted to demonstrate the method, one using data from Kew (Surrey), the other analysing the soil moisture balance at Bodiam (Sussex) for the same period, 1972–1973. Table I shows the calculation of the water balance at Kew. The soil was at capacity at the end of March. In April the potential transpiration exceeded rainfall by 14 mm; all this amount can be extracted at the full rate from the top layer of soil, in which the moisture content thus decreases from 50 to 36 mm. In May there is a potential loss of $78 - 24 = 54$ mm, only 36 mm of which can be taken at the full rate; the remaining 18 mm is at half rate, i.e., 9 mm from the second layer. In June all the potential loss of 72 is extracted at half rate, reducing the second layer content to 5 mm. In July a potential of 70 mm is taken out by 5 mm at half rate and 15 mm at quarter rate. In August all the remaining moisture in the third layer is extracted and the soil stays dry until November, when there is an excess of rain over transpiration of 50 mm, all of which is replaced in the top layer. This replacement process continues until the end of February, when 118 mm of soil moisture has been replaced, leaving a dry zone at the bottom of the third layer. There is no excess and drains do not run. The same process is then repeated in 1973 when soil moisture replenishment started in September.

The situation at Bodiam is shown in Table II. There are two main differences at Bodiam from the balance at Kew. Firstly, the available moisture in the soil was not fully exhausted during the summer of 1972. Secondly, the soil returned to capacity early in December and thereafter excess water, totalling 168 mm, drained through the soil until extraction began again in March. In April the soil again returned temporarily to capacity and a further small excess of 3 mm went to drainage.

It is important to remember that any excess rain during a summer month is stored in the uppermost layer of soil and is therefore available at the full potential rate during the next month. For example, having reached a soil moisture status stage of 0—0—25, if the next month had 120 mm rain and 90 mm potential transpiration, the status would become 30—0—25. If the next month had a 30-mm deficiency in rainfall, this would all be available from the top layer at full potential rate.

* The total water needed to restore capacity is the "soil moisture deficit".

TABLE I

Moisture balance at Kew, Surrey (in mm)

Month	Rainfall (IN)	Potent. transpiration (OUT)	Potent. gain (IN)	Soil gain (IN)	Excess (IN)	Potent. loss (OUT)	Full rate	Half rate	Quarter rate	Soil moisture status 0–50 mm	50–100 mm	100–125 mm
1972:												
Mar.	39	53	—	—	—	14	14	0	0	50	50	25
Apr.	24	78	—	—	—	54	36	9	0	36	50	25
May	17	89	—	—	—	72	0	36	0	0	41	25
June	20	90	—	—	—	70	0	5	15	0	5	25
July	10	88	—	—	—	78	0	0	10	0	0	10
Aug.	28	44	—	—	—	16	0	0	0	0	0	0
Sept.	15	22	—	—	—	7	0	0	0	0	0	0
Oct.			—	—	—	—	—	—	—	0	0	0
Nov.	56	6	50	50	0	—	—	—	—	50	0	0
Dec.	59	3	56	56	0	—	—	—	—	50	50	6
1973:												
Jan.	13	3	10	10	0	—	—	—	—	50	50	16
Feb.	13	11	2	2	0	—	—	—	—	50	50	18
Mar.	9	33	—	—	—	24	24	0	0	26	50	18
Apr.	47	55	—	—	—	8	8	0	0	18	50	18
May	60	75	—	—	—	15	15	0	0	3	50	18
June	67	110	—	—	—	43	3	20	0	0	30	18
July	54	96	—	—	—	42	0	21	0	0	9	18
Aug.	42	88	—	—	—	46	0	21	7	0	0	11
Sept.	71	55	16	16	0	—	—	9	—	16	0	11
Oct.	25	23	2	2	0	—	—	—	—	18	0	11
Nov.	29	6	23	23	0	—	—	—	—	41	0	11

TABLE II

Moisture balance at Bodiam, Sussex (in mm)

Month	Rain-fall	Potent. trans-piration	Potent. gain	Soil gain	Excess	Potent. loss	Full rate	Half rate	Quarter rate	Soil moisture status 0–50 mm	50–100 mm	100–125 mm
	IN	OUT		IN		OUT						
1972:												
Mar.	79	52	27	0	27	–	–	–	–	50	50	25
Apr.	56	78	–	–	–	22	22	0	0	50	50	25
May	49	89	–	–	–	40	28	6	0	28	50	25
June	50	90	–	–	–	40	0	20	0	0	44	25
July	23	83	–	–	–	60	0	24	3	0	24	25
Aug.	40	44	–	–	–	4	0	0	1	0	0	22
Sept.	24	22	2	2	–	–	–	–	–	2	0	21
Oct.	88	6	82	82	0	–	–	–	–	50	34	21
Nov.	116	0	116	20	96	–	–	–	–	50	50	25
1973:												
Jan.	35	3	32	0	32	–	–	–	–	50	50	25
Feb.	48	11	37	0	37	–	–	–	–	50	50	25
Mar.	19	33	–	–	–	14	14	0	0	36	50	25
Apr.	65	48	17	14	3	–	–	–	–	50	50	25
May	64	76	–	–	–	12	12	0	0	38	50	25
June	43	111	–	–	–	68	38	15	0	0	35	25
July	43	95	–	–	–	52	0	26	0	0	9	25
Aug.	24	86	–	–	–	62	0	9	11	0	0	14
Sept.	112	48	64	64	0	–	–	–	–	50	14	14
Oct.	37	24	13	13	0	–	–	–	–	50	27	14
Nov.	26	6	20	20	0	–	–	–	–	50	47	14

Daily balance sheets

The daily rainfall amounts are usually known, but to ascertain the daily transpiration is not so easy, because the Penman formula is not suitable for daily values, unless accurate radiation data is available. The first approximation would be to allocate the monthly values equally over the days, working in whole millimetres. For example, in April 1972 at Kew, the monthly value was 53, giving 1 mm for April 1—7, and 2 mm for April 8—30. A more detailed attempt would be to work in millimetres and tenths and to assume that the transpiration daily-values varied from the average of 1.8 mm by amounts proportionate to the sunshine experienced, which is quite a justifiable assumption. The following pattern was adopted: 12 h sun, 3.4 mm; 11 h, 3.2 mm; 10 h, 3.0 mm; and so on until 2 h sun, 1.4 mm; then 1 h, 1.1 mm, less than 1 h, 0.8 mm, no sun 0.6 mm. By this means, with very slight adjustments, the sum of the daily values equalled the monthly total.

The daily balance sheets for April 1972 are detailed in Table III. The monthly balance sheets gave soil-moisture deficits of 14 mm at Kew and nil at Bodiam. These more detailed analyses gave 17.5 mm and 4.6 mm, respectively. The differences are due to the more precise estimate of excess water which is drained away in the days when soil capacity is temporarily regained. They are most likely to occur early in the season; when the soil is well below capacity all the month, there should be no difference between monthly and daily balances.

If seasonal assessments are being made, an error of 5 mm is negligible, especially when there are certainly errors of this magnitude in measurements of rainfall, estimates of transpiration, and the empirical form of the model for actual transpiration.

This does not mean that a daily balance sheet is unnecessary. It may be necessary to know the moisture content of the topmost layer of soil which contains the first 10 mm of soil moisture, which is why Table III shows the soil moisture status in 2 columns, 0—10 mm and 10—50 mm. This may be important for a variety of purposes, such as germination or soil-parasite development.

Despite the differences in the monthly rainfall totals (39 mm at Kew, 72 mm at Bodiam) there was very little difference in the number of days when there was none of the first 10 mm of moisture available, 9 days at Kew and 7 days at Bodiam. On the other hand, there was only 1 day when the Kew soil was at capacity, but this state occurred on 11 days at Bodiam.

Evaporation from bare soil

Evaporation from bare soil is very difficult to estimate on a monthly basis because so much depends on the distribution of rainfall. Soil which is wet on the surface, during and just after rain, is losing water by evaporation and

TABLE III

Daily soil moisture balance sheets, Kew and Bodiam, April, 1972 (in mm)

Date	Rain	P.T.	Gain	Loss	Soil status 0–10	10–50 mm	Rain	P.T.	Gain	Loss	Soil status 0–10	10–50 mm
	Kew						*Bodiam*					
1	0.1	0.8		0.7	9.3	40	–	0.6		0.6	9.4	40
2	–	0.8		0.8	8.5	40	0.8	0.6	0.2		9.6	40
3	0.5	2.6		2.1	6.4	40	3.8	2.0	1.8		10	40
4	2.1	1.4	0.7		7.1	40	5.6	1.1	4.5		10	40
5	2.6	2.0	0.6		7.7	40	9.2	1.8	7.4		10	40
6	0.8	2.6		1.8	5.9	40	5.7	3.0	2.7		10	40
7	0.6	1.1		0.5	5.4	40	2.7	2.0	0.7		10	40
8	–	2.0		2.0	3.4	40	–	2.4		2.4	7.6	40
9	2.7	2.2	0.5		3.9	40	6.7	1.4	5.3		10	40
10	2.6	2.0	0.6		4.5	40	8.4	1.6	6.8		10	40
11	4.1	1.4	2.7		7.2	40	3.2	0.8	2.4		10	40
12	1.5	0.8	0.7		7.9	40	0.8	0.8			10	40
13	8.6	2.2	4.3		10	40	4.2	2.0	2.2		10	40
14	0.4	0.8		0.4	9.6	40	0.9	0.8	0.1		10	40
15	0.3	2.6		2.3	7.3	40	–	2.8		2.8	7.2	40
16	–	1.8		1.8	5.5	40	–	3.0		3.0	4.2	40
17	0.1	1.8		1.7	3.8	40	0.4	1.8		1.4	2.8	40
18	–	1.4		1.4	2.4	40	0.5	1.1		0.6	2.2	40
19	0.2	2.6		2.4	0	40	3.2	1.8	1.4		3.6	40
20	–	1.8		1.8	0	38.2	–	1.6		1.6	2.0	40
21	0.1	2.6		2.5	0	35.7	–	2.4		2.4	0	39.6
22	–	1.8		1.8	0	33.9	–	2.0		2.0	0	37.6
23	–	0.8		0.8	0	33.1	–	2.6		2.6	0	35.0
24	–	1.2		1.2	0	31.9	–	1.4		1.4	0	33.6
25	–	3.6		3.6	0	28.3	–	3.4		3.4	0	30.2
26	–	0.8		0.8	0	27.5	0.1	0.8		0.7	0	29.5
27	–	1.2		1.2	0	26.3	–	0.8		0.8	0	28.7
28	7.0	2.1	4.9		4.9	26.3	6.6	2.4	4.2		4.2	28.7
29	5.5	0.8	4.7		9.6	26.3	15.3	0.6	14.7		10	37.6
30	–	3.4		3.4	6.2	26.3	0.4	2.6		2.2	7.8	37.6

acting like a free-water surface. This rate of evaporative loss decreases rapidly as a dry-soil layer forms a barrier to the evaporative process; the soil tends therefore to seal itself against large continual losses.

The way in which the evaporation decreases depends largely on the type of soil and the extent to which it has been cultivated. As an example, the following model could be adopted. Dividing the first 10 mm of soil moisture into 5 layers of 2 mm each, then: the first 2 mm is evaporated at the full potential evaporation rate (which is assumed to be 25% higher than the potential transpiration); the second 2 mm at two-thirds rate; the third 2 mm at one-third rate; the fourth 2 mm at one-fifth rate; the fifth 2 mm at one-tenth rate; and assuming that in a climate without prolonged dry spells, a loss of 10 mm is about the maximum.

Similar models could be devised using different assumptions, but this example would provide the analysis shown in Table IV for Kew, April 1972.

Normally daily rainfalls are measured at 09h00 and "thrown back" to the previous day. This is inconvenient because any rain falling during the night

TABLE IV

Soil moisture balance sheet for evaporation from bare soil (in mm)

Date	Rain	Potent. evapor.	Direct loss	Soil moisture loss					Soil moisture status				
				Full rate	2/3 rate	1/2 rate	1/5 rate	1/10 rate	0–2	2–4	4–6	6–8	8–10
1	1.0	1.0	1.0	—	—	—	—	—	2	2	2	2	2
2	0.1	1.0	0.1	0.9	0	0	0	0	1.1	2	2	2	2
3	0	3.2	0	1.1	1.4	0	0	0	0	0.6	2	2	2
4	0.5	1.5	0.5	0	0.6	0	0	0	0	0	2	2	2
5	2.1	2.5	2.1	0	0	0.1	0	0	0	0	1.9	2	2
6	2.6	3.2	2.6	0	0	0.2	0	0	0	0	1.7	2	2
7	0.8	1.4	0.8	0	0	0.2	0	0	0	0	1.5	2	2
8	0.6	2.5	0.6	0	0	0.6	0	0	0	0	0.9	2	2
9	0	2.7	0	0	0	0.9	0	0	0	0	0	2	2
10	2.7	2.5	2.5						0.2	0	0	2	2
11	2.6	1.7	1.7						1.1	0	0	2	2
12	4.1	1.0	1.0						2	2	0.2	2	2
13	1.5	2.7	1.5	1.2	0	0	0	0	0.8	2	0.2	2	2
14	8.6	1.0	1.0						2	2	2	2	2
15	0.4	3.2	0.4	2.0	0.6	0	0	0	0	1.4	2	2	2
16	0.3	2.2	0.3	0	1.2	0	0	0	0	0.2	2	2	2
17	0	2.2	0	0	0.2	0.5	0	0	0	0	1.5	2	2
18	0.1	1.7	0.1	0	0	0.5	0	0	0	0	1.0	2	2
19	0	3.2	0	0	0	1.0	0	0	0	0	0	2	2
20	0.2	2.2	0.2	0	0	0	0.4	0	0	0	0	1.6	2
21	0	3.2	0	0	0	0	0.6	0	0	0	0	1.0	2
22	0.1	2.2	0.1	0	0	0	0.4	0	0	0	0	0.6	2
23	0	1.0	0	0	0	0	0.2	0	0	0	0	0.4	2
24	0	1.5	0	0	0	0	0.3	0	0	0	0	0.1	2
25	0	4.5	0	0	0	0	0.1	0.4	0	0	0	0	1.6
26	0	1.0	0	0	0	0	0	0.1	0	0	0	0	1.5
27	0	1.5	0	0	0	0	0	0.1	0	0	0	0	1.4
28	0	2.6	0	0	0	0	0	0.3	0	0	0	0	1.1
29	7.0	1.0	1.0						2	2	2	0	1.1
30	5.5	4.2	4.2						2	2	2	1.3	1.1
Total	40	65.3	21.7	5.2	4.0	4.0	2.0	0.9					

can only be evaporated the following day. It is therefore more sensible to "throw forward" for the purposes of a daily analysis. The same reasoning applies to the daily-transpiration analysis but as the rate of transpiration decreases only slowly with increasing soil moisture deficit, the error can be ignored.

It is furthermore assumed that if the rainfall is less than the potential evaporation, this amount of water is evaporated at the full rate without altering the soil-moisture status — again not strictly true, but it has the merit of simplicity.

Adding up the evaporative loss deduced from this model, the total comes to 38 mm which is very close to the rainfall. This is confirmed by the fact that the soil-moisture status is very similar at the beginning and end of the month. If a similar calculation was carried out for the Bodiam figures, with a monthly rainfall of 79 mm, it would be found that the evaporation was about 45 mm with again little change in soil-moisture status.

The evaporation from bare soil appears to depend largely on the incident rainfall and its distribution. The decrease in soil moisture cannot be large because of the sealing-off effect of a top dry-soil insulating layer. Cultivations during the fallow period, disturbing this protection, will increase the loss and so will the growth of any weeds.

This exercise confirms the value of a period of fallow in any crop rotation in a semi-arid climate, especially if the soil is covered with a stubble or trash mulch. To conserve the soil moisture the surface evaporation must be reduced and all vegatative growth kept to a minimum.

Numerous other models can be devised in order to calculate the soil-moisture status, but for precise estimates it is important to take into account the time of rainfall. Rain falling during the night cannot be treated in the same way as rainfall during the day, especially if it is required to find the length of time when the soil surface remains wet. If dealing with the moisture conditions of seeds below the surface such precision may not be necessary. A model using 12-h rainfall totals could be as follows: *day-time rain* (09—21 h) is evaporated at full rate up to potential evaporation, any remainder being added to the soil; any deficiency from potential being extracted from the soil at the appropriate rate; *night-time rain* (21—09 h) is all added to the soil.

This will give total evaporation estimates slightly greater than the simpler model.

Verification

The direct verification of a soil-moisture balance sheet is not easy, because it would involve intricate and possibly inaccurate measurements of soil moisture. Indirect verification can be obtained by application of the theoretical results as explanation of biological facts.

One simple verification of the monthly balance sheet is in the estimation of the date of return to capacity. If this estimate corresponds to the time when drains begin to run or well-levels begin to rise then a degree of corroboration has been obtained.

Partial crop cover

Having decided upon models for the estimation of soil-moisture status under bare soil and under complete crop cover, it is now only a matter of simple arithmetic to obtain a result for partial crop cover by combining the two in their appropriate proportions.

If this is done it will be found that in the early stages, when bare soil predominates, the main factors influencing the water loss and hence the soil-moisture status, are the frequency with which appreciable amounts of rain occur and especially the length of time between effective rains.

At a later stage, when the crop covers 25—75% of the ground, the water loss will be found to be very close to the long-term average of potential transpiration because of an inbuilt compensation effect. In wet cloudy weather the current potential transpiration will be lower than average but more water will be lost from wet soil; in dry sunny weather the current potential transpiration will be greater, but less will be lost by evaporation from the dry soil.

At stages nearing complete 100% cover the dominating factor is the actual potential transpiration.

METEOROLOGICAL MEASUREMENTS

The research worker who uses meteorological data, or who makes weather measurements on his own account, must always be aware of the possible errors. To begin with, any measurement can only be as good as the instrument used; more than that, the type of instrument determines the meaning of the measurement. For example, a thermometer only shows the temperature of its sensing element, whether it is a liquid such as mercury in a glass bulb, the electrical resistance of a metal element or any other measuring device. The instrument may be subject to calibration errors or suffer from hysteresis, in that its reading will be affected by the conditions of the immediate past. It may have a quick or slow response to changing conditions; it may have been badly maintained, or it may have suffered damage. All these factors contribute to instrumental errors, only some of which are readily avoidable.

Another type of error is due to mistakes by the observer himself. It is all too easy and by no means unknown even with experienced professional observers to misread a scale on an instrument or a line in a set of tables. Such mistakes are very difficult to correct after the event, so that some procedure of immediate checking is invaluable. Checking long after the event, often by a second person, can lead to the suspicion of errors which cannot be verified at a late stage. Theoretically such errors should be avoidable, but complete perfection is rarely attained.

The most important source of error probably lies in the way in which the instruments are used and especially in regard to the manner in which they are exposed to the element they profess to measure. Meteorological factors are so interrelated that the change of any one of them often leads to changes in the others which may or may not be significant. To obtain a reliable measurement of any one demands a standardisation of method, otherwise slight alterations of size, height, position and aspect can have undefined and unmeasurable effects. Uniformity of treatment within a network or throughout an experiment is therefore a first essential of any monitoring system.

The fourth difficulty encountered in the search for perfection lies in the variability of the meteorological elements themselves. One measurement site will provide a single sample; an adjacent site, apparently identical, could give quite different results. The differences may not always be large, but they may be significant within the terms of reference of the investigation. It may not be possible, for practical reasons, to take enough samples, but at very least, the degree of reliability of a single sample must be understood.

Every effort has to be made to eliminate errors, because bad measurements are worse than no measurements at all. The ways in which errors can be reduced, or if irremovable, can be assessed, will vary from element to element and at times, from climate to climate. The one important thing is not to ignore them and to pretend that they do not exist.

The need for measurements

To recapitulate, there are four main sources of error in the making of meteorological measurements: (1) the errors of the instrument itself; (2) the errors of the observer; (3) the errors due to exposure of the instrument; and (4) the sampling errors. The aim of any measurement method or technique is to reduce these errors to a minimum.

If the record of past agricultural events is to be investigated, the research worker has no alternative but to make the best use possible of existing weather measurements and to interpolate and extrapolate to the best of his ability and to the extent permitted by such existing records. If a new assessment or experiment is planned, it might still be possible to use the same existing reporting networks. Generally, however, it is wise to make special measurements which have been carefully chosen to obtain the type of weather record appropriate to the investigation.

The size of the unit

Before starting any such measurements, preliminary planning must take into account the scale of the investigation, or in other words "What is the size of the unit?" The meteorological unit must be compatible with the biological unit; it is useless to use approximate data for one type of variable and data of refined accuracy for another. The only exception to this basic rule is when a dependent variable is very insensitive to changes in an independent variable, in which case the latter need only be measured to a lower scale of accuracy.

The first unit of importance is that of time; the scale which is of importance may be a period of days, single days, or a period of hours. Broadly speaking, work in periods of minutes or seconds enters the realm of micrometeorology and few experiments on a field scale need this degree of precision in time.

Insufficient scales of measurement can lead to frustration in subsequent analysis, because it may be found, all too late, that more detailed data should have been obtained. On the other hand, with modern instruments and data-logging systems, it is very easy to take too many measurements, with the result that the investigator is swamped with data. Even with the assistance of means of mechanical analysis, far too much time and money will be spent dealing with excessive amounts of data. Clear thinking at planning time would have obtained equally useful and valid results. Poverty in research finance may be a blessing in disguise, in that it forces the scientist to make the best use of his limited means and to think before acting.

The size of the unit depends on two main factors; the degree of variation of the parameter itself under field conditions and the requirements of the problem. These two factors may be incompatible and in any case, the needs of the problem are rarely completely known until the problem is almost solved. In practice, therefore, it is generally the nature of the parameter and its normal variations in time and place in the environment which determines the size of the unit, but this is influenced by the time, money and manpower available.

One further point must be borne in mind, namely that it generally takes much longer to analyse the data than is realized. If only a limited time is available for the completion of a project, it is useless to spend time on the making of observations which cannot subsequently be analysed within the schedule. When observations and measurements are required for the whole of a growing season, there is, of course, no alternative to the length of the observing period. It is often wise to conduct a series of pilot experiments first to establish the potential of a measurement system and then run the full experiment during the second year. This is often far better than trying to run the experiment for two years in an inappropriate manner.

When planning an experiment or investigation, it is of the utmost importance that the scientist should not relegate himself to the role of servant of a data-processing machine. He must retain independence of thought and use his powers of logic and reason. If the apparent advantages of punch cards, magnetic tapes, computer programmes and such modern aids to data hand-ling and analysis are over-valued, then their requirements and limitations become the over-riding factor. The result often is that the data is not fully explored by that supreme computer, the human brain, and the analysis is carried out only in a clumsy fashion, in which speed and convenience have taken undue precedence. Computers are excellent for accelerating the use of an established procedure, they are no substitute for thought.

AGRICULTURAL DATA

Meteorological data may be profuse, agricultural data are conspicuous by their relative scarcity. When statistics do exist, they are likely to be limited in accuracy and homogeneity. These limitations have to be accepted because data acquisition is an expensive and time-consuming process. Meteorology is extremely fortunate to have a regular observing network; other sciences have far less basic facilities.

Agricultural data comes in various shapes and sizes. National and regional details, on a yearly basis, may be available regarding crop acreage, animal populations and production statistics. With few exceptions, yield data are subject to error and should not be considered as more than a general guide to performance. In certain sections of the agricultural industry such as milk or wool, reliable information may be available in certain countries, but general-ly the oft-quoted jibe "lies, damned lies and statistics" must be borne in mind.

More reliable information is sometimes obtained as a result of special surveys over more limited areas and over shorter periods of continuous rec-ord. The main difficulty here is one of consistency. It is very difficult to set up such an investigation using a number of different observers of variable ability, experience and conscience, without introducing some degree of non-uniformity. The results of widespread field trials are often prejudiced by similar difficulties, although it is always hoped that in a large enough choice of sites, the unrecognizable "local" effect will tend to cancel out.

In a search for precision, the tendency will always be towards smaller and smaller scale experiments. This, unfortunately, introduces a further disturb-ing factor, in that it becomes more difficult to acquire a true measure of the physical environment. In plot trials, for example, randomized according to the best statistical requirements, each plot affects its neighbours in regard to the all-important micro-meteorological profiles, and the standard variation of the results can be confusing. The extreme case is that of pot trials with plants and here it is well-nigh impossible to use standard meteorological measurements. For any validity in such experimental work, it is necessary to employ the intricate and expensive techniques of micro-meteorology, other-wise only the broadest of weather-growth deductions can be made.

However difficult plant investigations may be, it is true to say that animal experiments are even more fraught with possible error. Animals react to many unquantifiable factors such as previous experience, method of handling, speed of adjustment to change in circumstances and innate independance in action and behaviour. An attempt to set up a controlled experiment may lead to almost complete lack of control and the sheer artificiality of the circumstances renders its results useless for subsequent application. Any small-scale results must be verified by field-scale trials; unconfirmed extrapolation can be dangerous. Conversely, any result obtained from generalized data on a field-scale may not apply to individual trial plots or small groups of animals.

Meteorological data is thus not only more readily available and more profuse, but it is generally of a more reliable quality than biological data. Nevertheless, the only practical solution is to recognize the fallibility of the working material and to proceed with a realization of its inherent limitations.

Any plans to record agricultural events for a particular investigation must be carefully thought out. Although it is often difficult to attain a uniform standard of observation when several workers at different sites are involved, special attention must be paid to phenological or crop-growth stages. To try to analyse information on a simple calendar scale is to handicap the investigation from the start. Furthermore, in the same way in which meteorologists should be the best observers of weather data, agriculturalists or biologists should be always the more capable in the observation of agricultural data. This is yet another example of circumstances where co-operation between the disciplines is favourable to success.

The guiding rule is for the meteorologist to consult the experienced biological scientist at all stages, and especially at the start of any project. It is stupid to spend a long time working with data simply because it is available without first finding out to what extent such data can be trusted. The publication of data in any science or industry does not invariably imply an inherent accuracy.

SOURCES OF REFERENCE

It is unwise to start any new line of research, without first taking note of any previous work in the subject. Publication is no guarantee of accuracy but nevertheless it is still worthwhile spending considerable time on reviewing a subject before starting.

The best review papers on subjects in agricultural meteorology are those published by the World Meteorological Organization in Geneva in their series of *Technical Notes*. Such publications are prepared by working groups or individual authors, representing the best available sources of knowledge on each particular subject, and the coverage is international; most of such *Technical Notes* contain a valuable list of references for further study. A descriptive list of these publications is given in Appendix I. Even if this technical Agency of the United Nations has not produced any such review on a particular subject, it is advisable to approach their Commission for Agricultural Meteorology, through W.M.O. Geneva, for help and guidance regarding previous work. It is possible, in this manner, to be put in touch with a world authority on a subject or with a research station which is currently engaged in relevant research.

Each issue of the journal *Agricultural Meteorology* (Elsevier, Amsterdam) contains some 500 titles of recently published papers from an extensive variety of international journals. Titles alone are given, so that the precise contents of each paper have to be ascertained from an abstracting service, or by reading the appropriate journal or by obtaining a reprint. Titles are given in English.

Titles and abstracts of published papers in the subject are published annually by the German Federal Republic Weather Service in Frankfurt (edited by M. Schneider). This is a unique publication of its type, although the coverage is chiefly West European. The main language used is, naturally, German.

Some countries, through their meteorological services, issue annotated bibliographies of works in agricultural meteorology at regular intervals. These are circulated through international meteorological channels. Such lists do not claim to be comprehensive and include only work done in the country of issue, but they probably contain the most important references from the period under review. Canada, Israel, Switzerland, Austria, Finland, The Netherlands and Poland are some of the countries providing such information.

The number of publications dealing exclusively with agricultural meteorology is very limited, and the vast majority of research results are to be found in agricultural journals. Any search must therefore include consideration of such sources. The field is a vast one, but work is made considerably easier by reference to the abstract journals published by the Commonwealth Agricultural Bureaux, whose headquarters are at Farnham Royal, Slough, Buckinghamshire SL2 3BN, England.

The C.A.B. produce a variety of abstracts with a very wide international coverage of scientific journals. Written in English, their references indicate the original language of the published papers, together with the languages used in the paper summaries. Their abstracts give an excellent indication of the contents of each paper.

The C.A.B. consists of a series of abstracting bureaux. One, the Commonwealth Bureau of Pastures and Field Crops is situated at the Grassland Research Station, Hurley, Berkshire SL6 5LR, and issues monthly two journals: *Field Crop Abstracts*, which has a special section on agro-meteorology and *Herbage Abstracts* which also has an agricultural meteorology and climatology section. References to meteorological aspects can also be found in papers under other section headings. For example, in *Herbage Abstracts*, there are sections on irrigation, plant—water relations and environmental contamination.

Horticultural Abstracts is prepared by the Commonwealth Bureau of Horticultural and Plantation Crops at the East Malling Research Station, East Malling, Kent, England. Sections on climatic effects are contained under the following headings: general aspects of research and its application; temperate tree fruits and nuts; apples; vines; vegetables; beans; tomatoes; sub-tropical fruits and plantation crops.

Forestry Abstracts is issued by the Commonwealth Forestry Bureau, Oxford, England, and the most productive sections are: II. Site Factors; III. The Atmosphere, Meteorology, Climate and Microclimate; 18. Plant Ecology; 181.2. Atmospheric Relation, Acclimatization; 181.3. Water, Soil and Root Relations; 181.8. Phenology; 26. Combination of Forestry with Agricultural and Pastoral Husbandry (including shelterbelts); 4. Frost Injuries and Protection; 43. Forest Fires.

Abstracts dealing with Soils and Fertilizers are issued by the Commonwealth Bureau of Soils, Rothamsted, Hertfordshire, England. Therein the most appropriate sections are: 132. Soil Water; 136. Soil Temperature; 18. Soil Formation, Erosion; 21. Cultivation; 7. Climate, Geomorphology, Geology; 267. Irrigation; 268. Drainage.

. Other C.A.B. publications which may contain reference to agricultural meteorology topics are: *Review of Applied Entomology; Review of Plant Pathology; Animal Breeding Abstracts; Veterinary Bulletin; Dairy Science Abstracts; Weed Abstracts.*

One particular C.A.B. service of value is the publication of annotated bibliographies on special subjects; covering all issued papers over a series of years. Lists of typical subjects dealt with are given in Appendix II.

Further valuable sources of information are also to be found in the reports of the proccedings of international meetings sponsored by the United Nations Technical Agencies, W.M.O., F.A.O. and Unesco. In particular, attention should be paid to the series of training seminars conducted by the World Meteorological Organization, and to the various symposia reported in full detail in Unesco publications (see below).

Non-governmental international societies, such as the International Society of Biometeorology, also hold regular congresses at which many valuable papers are presented (see below). This society is at present engaged on the publication of a series of volumes entitled *Progress in Biometeorology*, with a large variety of contributors, the Editor-in-Chief being their Secretary, Dr. S.W. Tromp (Biometeorological Research Centre, Oegstgeest, The Netherlands).

W.M.O. Training Seminars

No date: Final Report. International Seminar on Tropical Agro-meteorology. Maracay, Venezuela, Sept. 1960. Servicio de Meteorologia y
Comunicaciones fav Venezuela, 213 pp.

1968:   Proccedings W.M.O. Regional Training Seminar in Agricultural
Meteorology, Wageningen, May 1968. Agricultural University,
Wageningen, 425 pp.

1968:   Agricultural Meteorology. Proceedings W.M.O. Seminar, Mel-
bourne, Dec. 1966. Bureau of Meteorology, Australia, Vols. 1 and
2, 733 pp.

1972:   Agricultural Meteorology. Proceedings W.M.O. Seminar with Special Reference to Tropical Areas, Barbados, Nov. 1970. W.M.O.
Geneva, Publ. No. 310, 357 pp.

Publications by Unesco, Paris

1963:   Changes of Climate. Proceedings Symposium Rome, Oct. 1961,
Arid Zone Research, XX, 488 pp.

1965:   Methodology of Plant Eco-Physiology. Proceedings Symposium
Montpelier, April, 1962. Arid Zone Research, XXV, 531 pp.

1968:   Functioning of Terrestrial Ecosystems at the Primary Production
Level. Proceedings Symposium Copenhagen, July 1965. Natural
Resources Research, V, 516 pp.

1968:   Agroclimatological Methods. Proceedings Symposium Reading,
July 1966. Natural Resources Research, VII, 392 pp.

1973:   Plant Response to Climatic Factors. Proceedings Symposium Upp-
sala, Sept. 1970. Ecology and Conservation, V, 574 pp.

Publications by International Society of Biometeorology

1972:   Biometeorology. Proceedings of the 6th International Biometeorological Congress Noordwijk, Sept. 1972. Volume 5, Part I, 153 pp;
Volume 5, Part II, 271 pp. Swets and Zeitlinger BV, Amsterdam.

Previous congresses of this society, and the publishers of their proceedings
are as follows: 1st Vienna, Sept. 1957. Johnson Reprint Corporation, New
York, N.Y. 2nd London, Sept. 1960. Pergamon Press, Oxford. 3rd Pau, Sept.
1963. Pergamon Press, Oxford. 4th New Brunswick, Aug. 1966. Swets and
Zeitlinger, Amsterdam. 5th Montreux, Sept. 1969. Swets and Zeitlinger,
Amsterdam.

CHOICE OF SUBJECT

Two main factors governing the choice of subject are the ability and experience of the scientists concerned and the amount of basic data with which they can work. Synthetic experience can be gained by reading published papers on the subject or consulting others with the necessary experience; data can be acquired, with the appropriate time lag, if sufficient manpower and money are available; there is no substitute for ability. Even so, limited ability can be put to maximum use if it is supplemented by an interest in the work and a desire to provide a useful service, provided that it is given adequate opportunity and support.

One further point should be borne in mind: the most useful research in agro-meteorology is that which leads to increased knowledge which can be put to practical use, either operationally on a tactical basis, or in planning, which involves agro-climatology. The discovery of a crop—weather or animal—weather relationship which cannot be applied in agricultural decision making is of limited value, adding only to the personal satisfaction or professional esteem of the person or persons concerned. At the very most it is an act of faith in the hope that someone, sometime, somewhere, will find a non-academic use for it. It is true that this hope is not always a vain one, but if priorities have to be specified, the decision must always rest on the answer to two questions: Can it be done? Is it worth doing?

Planning problems in agriculture

As stated above, planning criteria are the domain of agro-climatology and fall into two categories; the design or choice appropriate to a given site and the selection of a site for a given plan or intent. Leaving aside for the moment the problem of climatic stability which will be discussed in another section, it is essential that these criteria have been established as valid. Such criteria can be found by two methods of approach. In agro-climatology the aim would be to relate known states of site design or site selection to the known climate; herein the climate refers to past events and is dependable, limited only by the inadequacy of the parameters and the choice thereof, it is assumed that existing solutions of agricultural problems have led to a satisfactory state of choice of design, proven by years of trial and error. In other words, if the practice was not valid, it would not have persisted; this is not always true.

The alternative approach is to bring together all the established agro-meteorological criteria pertinent to the particular problem and so define the limits of acceptable climate. The main difficulty is that such criteria may not all be available. Furthermore, as has been emphasized elsewhere, empirical criteria may be a very dangerous import and a relationship which works excellently in one climate may be totally useless or even misleading elsewhere. Nevertheless, with care, the composite picture of climatic requirements and danger thresholds can be built up and then the agro-climatological investigation can proceed. The duty of the agro-meteorologist is to supply the climatologist with as many reliable criteria as possible. All aspects of the application of meteorology to agriculture are involved in agro-climatology both in site design and site selection. The whole complexity of individual and inter-related problems fall within the category of "land use". This is a huge strategical exercise and involves large sums of capital expenditure, both personal and national. The maintenance of man's heritage in the land depends on the correct solution to its problems. Therefore, the abovementioned duty of agro-meteorology has implications far beyond assistance to day-to-day decision making. The long-term decisions are of fundamental importance and must involve the correct interpretation of climate.

Certain aspects of applied meteorology in the agricultural industry are wholly connected with climatology. These are those which demand no operational decisions, for the farmer has no radius of action. Examples of this are drainage and liability to storm damage. A farmer can drain his land in accordance with the climatic risks, but once the adverse soil-moisture conditions occur there is nothing further he can do about it. A farm can be situated in an area least likely to encounter gales or hailstorms, but if either occurs there is nothing further that can be done. Livestock, it is true, can be moved out of an area liable to flood, but a crop cannot be moved from the same fields; harvesting is not an instant procedure.

A farmer may have some control over the creation of pollution, but he has no options if he is at the receiving end, especially in regard to air-borne pollution. His only chance is to select a relatively clean air area or select a crop which will endure the threat.

Many of the factors which affect the growth and development of crops such as rainfall and sunshine are beyond human control. Again, therefore, the options are climatically controlled: the choice of site or the choice of crop. Once the decisions have been made, once the crop has been sown, the farmer is at the mercy of the significant weather. If a crop is ripe it must be harvested as soon as possible; the weather cannot be manipulated.

Much the same degree of operational impotence exists in regard to frost incidence. Although some protective measures do provide a partial (and possibly expensive) remedy, the best defence is still one depending on climate, namely the avoidance of areas liable to frost.

Tactics and day-to-day decisions

Many of these decisions are highly sensitive to the current weather and the weather of the immediate past. Most of these can be grouped together under the heading of the timing of husbandry operations. The timings under maximum personal control are those concerned with action against pests and diseases of crops and animals. Similar, but less complete choice, exists in regard to soil cultivation and sowing.

Two major aids to crop growth, irrigation and fertilizer application, are under control to a great extent; additions can be made but subtractions are well-nigh impossible.

Other aspects of agriculture involve minor tactical weather-based decisions but major climatological design and choice decisions, namely, shelter, shade, use of glass or plastic cover and anti-erosion methods. The design and provision of housing and storage facilities also come into this category, for the only operational decisions are probably those of heating or ventilation.

Major operational decisions with a degree of weather sensitivity exist in the cases of animal feeding and animal reproduction, and other weather affected tactics are used in regard to the application of mulches and measures against forest or bush fires.

Seed germination is largely dependent on conditions following sowing and the possibilities of aid to pollination are limited.

The choice of crop variety or breed of animal cannot be based on weather considerations until good seasonal or climatic forecasts are available. Even if variety trials can be analysed with respect to the weather, the application of such results can only be on a climatological basis at present.

The major weather influences, with therefore the best potential economic record for applied meteorology and climatology are thus:

Operational	*Site design*	*Site choice*
Pests and diseases	Shelter	Frost
Irrigation	Housing	Land use
Husbandry timing	Erosion	
Fertilizer application	Land use	

All other agricultural operations and decisions have a weather factor with variable but generally lesser degrees of significance and importance.

SUGGESTIONS FOR FURTHER READING

Meteorological observations and measurements

Hogg, W.H., 1949. Effect of heated glasshouses on minimum temperatures in a nearby screen. *Meteorol. Mag., Lond.,* 78: 335—336.

Gloyne, R.W., 1950. An examination of some observations of soil temperatures. *J. Brit. Grassld. Soc.,* 5 (2): 157—177.

Gloyne, R.W. and Smith, L.P., 1951. Shielded thermometer mounts. *Meteorol. Mag., Lond.,* 80: 203—204.

Smith, L.P., 1951. Random errors in standard observations. *Meteorol. Mag., Lond.,* 80: p. 236.

Gloyne, R.W., 1953. Radiation minimum temperature over a grass surface and over a bare soil surface. *Meteorol. Mag., Lond.,* 82: 263—267.

Gloyne, R.W., 1965. The "standard" climate — its measurement and significance for vegetation. *Sci. Hort.*, 17: 137—150.

Gloyne, R.W., 1965. Some empirical relationships concerning the intensity of direct solar radiation. *Meteorol. Mag., Lond.*, 94: 401—410.

Hogg, W.H., 1968. Network planning, observations and instrumentation. *Proc. W.M.O. Reg. Sem. Agromet., May, 1968, Wageningen*, pp. 275—288.

Gloyne, R.W., 1971. A note on the measurement and estimation of evaporation. *Meteorol. Mag., Lond.*, 100: 322—328.

Cochrane, J., 1972. A method for checking the validity of earth temperature data. *W.M.O. Publ.*, 310: 219—222.

Derived and estimated data

Gloyne, R.W., 1952. Daily maximum temperature on the surface of the ground. *Meteorol. Mag., Lond.*, 81: 203—206.

Smith, L.P., 1960. Formulae for average sunshine values. *NAAS Q. Rev.*, 11: 175—178.

Taylor, S.M. and Smith, L.P., 1961. Estimation of averages of radiation and illumination. *Meteorol. Mag., Lond.*, 90: 289—294.

Armour, D.G., Balloch, J. and Gloyne, R.W., 1971. The estimation of mean dry-bulb temperatures during daylight hours. *Meteorol. Mag., Lond.*, 100: 220—222.

Potential transpiration and soil moisture balance

Anonymous, 1954. The calculation of irrigation need. *U.K. Minist. Agric. Fish. Food Tech. Bull.*, 4: 37 pp.

Anonymous, 1967. Potential transpiration. *U.K. Minist. Agric. Fish. Food Tech. Bull.*, 16: 77 pp.

Anonymous, 1971. The significance of winter rainfall over farmland in England and Wales. *U.K. Minist. Agric. Fish. Food Tech. Bull.*, 24: 69 pp.

CHAPTER 3

The Modes of Agricultural Meteorology — Plants

> "He that can find out an improvement in husbandry
> doeth more good to the commonwealth than he that
> founds fifty almshouses."
>
> H. PLATTES

> "He gave it for his opinion, that whoever could
> make two ears of corn or two blades of grass to
> grow upon a spot of ground where only one grew
> before, would deserve better of mankind, and do
> more essential service to his country than the
> whole race of politicians put together."
>
> JONATHAN SWIFT

SEED GERMINATION

The germination of seeds is affected by many factors which can be sum-marised in the following manner.

(a) The nature of the seed; its requirements for warmth and moisture. These requirements can be materially affected by the provenance of the seed, the meteorological factors during its formation during the previous year, and the conditions of its storage.

(b) The temperature of the soil surrounding the seed; the supply of heat energy to the seed by conduction from its surroundings. These conditions are influenced by the nature of the soil, its radiation balance at the surface, its heat conductivity, its moisture content and its water table.

(c) The availability of moisture in the soil surrounding the seed. This is influenced by the nature of the soil, the water balance of the soil surface, the diffusivity and infiltration of water within the soil, the height of the water table and the drainage.

(d) The nature of the soil and soil surface, which affects (b) and (c) above. The results of experiments in soil will not necessarily agree completely with those carried out in solutions in the laboratory.

(e) The external weather, especially radiation and rainfall. Experiments in plots in the open may give different results from those in pots in a green-house or glasshouse or artificial climate-control chamber.

(f) The presence of adverse influences, which may be pathological, ento-mological or animal. These disadvantages may be combatted by soil steriliza-tion and seed dressings.

For a seed to be able to germinate, it has to absorb moisture. The ease with which it can do this is determined by the temperature and the available water. In general it may be said that there exist minimum temperature and moisture levels in the soil, but maximum levels have also to be considered. Soil temperatures at seed depth, even in temperate latitudes can be too high for gemination in mid-summer. Soils saturated with water beyond the state known as field capacity can contain too much water for efficient germination, in addition to the fact that such conditions depress soil temperatures. It follows that adequate drainage is of great importance in early spring when rainfall often exceeds evaporation and there are liable to be periods when the soil contains an excess of water which must be drained away as quickly as possible. Later in the year, with greater strength of solar radiation and more rapid evaporation, the moisture content of soil surrounding shallow-sown seeds can often be reduced below the necessary minimum levels and consequently frequent additions of irrigation are necessary to ensure rapid germination. It is the frequency which is important. Twice the amount of added water will not keep the top layer of soil moist for twice the time, it will merely add water to the soil layers below seed depth.

The effect of temperature

The relation between germination and subsequent root growth, soil-moisture conditions not being limiting, and soil temperature, is not linear. Moreover, although maximum and minimum limiting temperatures are often quoted with reference to the requirements of various seeds, it is probable that there is no precise cut-off level and that the graph of activity against temperature becomes asymptotic at the so-called critical temperature values.

Within these limits, there is an increase of activity up to an optimum temperature level, thereafter a more rapid decrease with further increase of temperature to the maximum. The shape of the graph therefore resembles the track of a somewhat lofted drive at golf against a head-wind; some 2/3 to 3/4 of the drive covers the period when the golf ball is rising, during the remainder it is falling rapidly.

Thus, although the growth—temperature curve is not linear the part of the curve between the effective minimum temperature and the optimum approximates to a straight line, so that a useful meteorological parameter is the accumulated temperature in day-degrees, measured above the minimum temperature for emergences. This temperature—time factor (often wrongly called a "heat sum") is most useful in this context, but it should not be used indiscriminately as a parameter for further stages of crop development and yield, even when moisture conditions are constant, because it ignores the important radiation effects except in so far as soil temperature itself is a manifestation of radiation balance. Also it must be emphasized that the temperature is that of the soil at seed-depth; air temperatures are of very limited use in consideration of germination problems, although links obviously exist between air and soil temperatures.

A difficulty arises in that there is a large diurnal variation in soil temperature at shallow depth, so that once-daily readings of temperature are of limited value. For precise experimental work a constant record of temperature is essential and graphical methods of temperature-time summations or acceptable mathematical integration formulae must be used to obtain accurate results. For practical purposes in the field it is often possible to refer the precise measurements at seed depth to a reference observation taken once a day at 30 cm depth where the diurnal range is very much smaller. In this way it is possible to translate experimental results into a form in which they can be used operationally, but in this process a degree of uncertainty is bound to be introduced. This is inevitable, as simplification generally introduces lack of precision, but the important thing is to be aware of the magnitude of the resultant error and to ascertain whether it is of practical significance.

If a constantly operating thermograph cannot be used in the soil, a maximum—minimum type of soil thermometer can provide helpful results. The simple mean of maximum and minimum can be taken to represent the mean 24-h temperature with no great loss of accuracy, provided that each lies within the approximately linear part of the growth—temperature curve; if the day maximum exceeds the optimum but not the germinating maximum, an error will be produced; if either day maximum or day minimum exceeds the germination maximum or minimum, the error will be considerable.

The effect of soil moisture

The effect of soil moisture, or more strictly the osmotic pressure, is very complicated. Seeds vary according to their structure and especially in regard to the nature of their outer surfaces. It is thought that they need to take up their own weight in water for satisfactory germination and that they are unaffected by water conditions in the soil distant some 5 times their own size. The effects of soil moisture cannot be separated from those of soil temperatures because the ability of seeds to assimilate moisture depends on the temperature conditions.

The existence of adequate soil-moisture conditions is most important in the early days after sowing. Once roots have begun to form and develop they can tap water resources at lower depths and maintain the process of germination and shoot development even though the seed itself may then be in a layer of soil which has experienced considerable loss of moisture by direct evaporation from the soil surface.

Generally speaking, however, the less the soil moisture is available and the greater the osmotic or matrix pressures, the greater is the need for higher temperatures or longer time for germination, that is, the higher the accumulated day-degrees. Soils can be too wet for germination and emergence, so that excessive natural rainfall or excessive watering can impede growth irrespective of soil temperature.

For rapid efficient germination, both a high soil temperature and an adequate soil-moisture supply are necessary. In field conditions, however, these two requirements are in opposition, they demand contradictory weather in a temperate climate. Warm soil will occur in dry weather, cooler soil in years with frequent rain or showers, although occasionally a mild spring is also a wet one, at least over short periods of days.

It is very difficult, especially in the field, to maintain accurate records of soil moisture. Therefore, experiments dealing with the effect of physical conditions on germination success are well advised to concentrate first on the effects of soil temperatures, and to take steps, by appropriate watering, to see that soil-moisture conditions are kept in a non-limiting and approximately constant state.

POLLINATION

The weather at the time of pollination is a very critical factor in the life cycle of all plants. Even in cultivated crops, especially those grown for seed and for fruit and some vegetables, the extent of succes in pollination may be decisive for the final yield.

Much pollination is carried out by insects such as honey-bees, or it can be wind-born without intermediary aid, but in all cases the weather will determine the amount of pollen, the time when it is available and the extent to which it is transmitted to other plants.

Experiments on the effects of weather on pollination are difficult to plan and conduct, because of the precise nature of the phenological stage of the plant and the critical nature of the weather. Warmth is generally a beneficial factor; rain, especially in the middle of the day, can be counter-productive. Wind may help air-borne dissemination, but handicap the activity of insects. Insects, being cold-blooded, react quickly to conditions favourable to heat loss, so that a bee will rarely work below a certain minimum temperature and will be able to carry out more flights under conditions of a light wind.

In this context, it is probable that one of the reasons why fruit orchards need shelter from the wind is that it provides better pollination and set of the fruit. Recent experiments have also suggested that orchard heating during flowering can not only offset a degree of frost at night, but also raise the day-time temperatures and encourage the activity of pollinating insects.

An indirect proof of the role of temperature can be obtained by considering the yields of honey from professional bee-keepers. Work on this problem was carried out by Hurst in 1967, and as an interesting example of serendipity (the process of finding something that was not being sought), disclosed a potentially useful indicator for seasonal weather forecasting (Hurst, 1970). The Honey Producers Association provided data on the average yield of honey per hive over Great Britain for 1928—68. Taking the 8 years with the 8 highest honey yields and the 8 with the worst, Table V shows the relation-

TABLE V

Honey yields and summer temperatures

	Years of good honey	Years of poor honey
Mean honey yield (lb/hive)	76.9	12.6
Mean days with temperature above 20° C — July	23.7	11.7
— August	23.3	13.9
Mean days with temperature above 25° C — July	9.6	1.5
— August	6.5	1.9

The variation about the means in all cases was low.

ship with the warmth of the days in July and August, using Kew Observatory as a reference station.

It was now noticed that the years of good honey yields were those when there was an early warming-up in the stratosphere, coupled with a reversal in wind direction, which takes place in late winter or early spring. The years 1952—1968 were therefore examined in greater detail, shown in Table VI. Two inferences can be drawn: firstly warm weather in July and August favours honey production; secondly early stratospheric warming is a good indicator of a warm summer in the British Isles (see Fig. 1).

Pollen distribution is an essential process for plant reproduction, but may be a disadvantage to human beings. For example, perhaps some 10% are allergic to some degree to the pollen of grasses, in that they suffer from the disease known as "hay-fever". At St Mary's Hospital in London, with the aid of a spore trap, accurate records have been kept for over 10 years of the pollen concentrations in summer. Analysis of this unique series of data has shown that the weather of April and May effectively determines the amount of grass pollen likely to be released and also the earliest date when a concentration could be expected which will give rise to serious disease symptoms. The selected weather parameter was the sum of the mean April and May temperatures (T_4) averaged over 4 reporting stations surrounding London. The results are shown in Table VII.

The forecast of the severity of the hay-fever season (see Fig. 2A), based on the total pollen count is satisfactory. The forecast of the start of the season (see Fig. 2B) is less reliable, because it is likely to determine the first available date, which in itself may be correct for the state of pollen, but cannot take into account the weather of the day itself which will determine the actual pollen release. The previous temperatures therefore indicate the potential start; the actual start depends on the weather at the time.

Nevertheless, using only previous meteorological data it is clearly possible

TABLE VI

Stratospheric warming, summer temperatures, and honey yields (lb./hive)

Year	Honey yield	Days with 20° C (July, Aug.)	Days with 25° C (July, Aug.)
Years of early stratosphere warming			
1952	30	39	8
1955	75	48	18
1957	55	28	9
1959	65	51	20
1961	54	33	3
1964	35	46	11
Mean	52	41	12
Years of average stratosphere warming			
1960	21	20	0
1963	16	23	4
1967	49	49	10
Mean	27	31	5
Years of late stratosphere warming			
1953	15	31	3
1954	10	14	0
1956	20	15	4
1958	10	30	3
1962	18	30	0
1965	10	28	0
1966	19	29	4
1968	22	30	4
Mean	15	26	2

to warn hay-fever sufferers and their medical advisors, in regard to the prospects of the coming summer and this itself is a considerable benefit. It is further more possible to analyse the subsequent weather, day by day, to obtain a more precise relationship between current weather and the amount of air-borne pollen.

This is not a simple process. It is first necessary to adjust each year to the same starting date and also to make the pollen amounts comparable by dealing with percentages of the total seasonal count. Next it must be realized that London is not surrounded by equal amounts of grass on all sides so that an adjustment must be made for wind direction in accordance with grassland distribution.

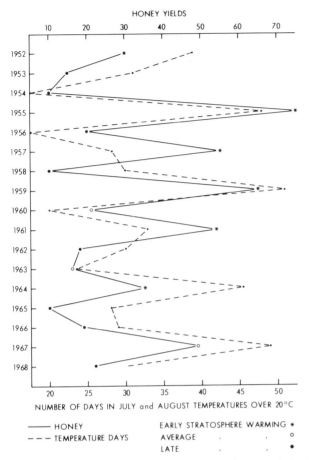

Fig. 1. Honey yields, late summer temperatures and stratospheric warming.

TABLE VII

April and May temperatures and grass pollen counts in London

Year	T_4 (°C)	Sum of mean daily pollen counts (× 1,000)	First day of significant count
1964	22.4	85	June 5
1961	21.4	64	June 4
1966	19.3	50	June 7
1965	20.1	42	June 11
1962	18.4	41	June 10
1969	19.6	40	June 8
1967	18.7	35	June 12
1963	19.4	31	June 10
1970	19.9	29	June 8
1968	18.3	27	June 9

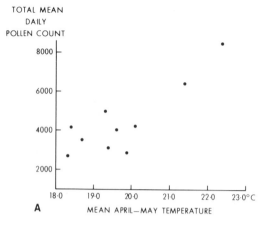

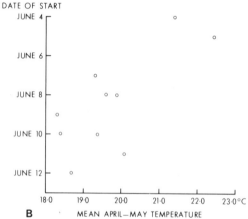

Fig. 2. A. Seasonal pollen counts and late spring temperatures. B. Start of hay fever symptoms and late spring temperatures.

The daily data for each year is now homogeneous but it cannot be assumed that pollen is equally available for dispersal throughout the season. However, by simple averaging of the adjusted data and ignoring the daily weather, an approximation of the pollen release curve can be obtained. The data can then be further adjusted to take into account the pollen release phase.

It is now possible to analyse the data with respect to the weather factors which are relevant to the sample catch, namely: rainfall, temperature, sunshine, wind force and turbulence, atmospheric instability and convection. This having been done, the pollen release curve can be re-determined with full adjustment for current weather. This release curve can either be described in terms of days after the start, or in terms of the proportion of total pollen that already has been released.

Linearity of weather—pollen relationships cannot be assumed. For example, day-time rainfall below 2 mm apprears to have no effect and above this value pollen catches are reduced by approximately one half, although the proportion depends on the timing and duration of the rain. No pollen is caught if the day maximum temperature is below 14°C and above 22°C the rate of increase with temperature appears to decrease. Pollen catches increase with sunshine hours and decrease with wind strength and with instability.

With the results of such analyses, a model of pollen catch can be devised which will provide a guide to days of severe hay-fever.

Data of this type lends itself to multi-regression computer-based analysis, but care must be taken to ensure the homogeneity of the values used. In other words, a great deal of thought must be given before using any sophisticated machine aids. A mathematical analysis will always give some kind of answer, but it is incapable of giving a sensible answer unless it is given reliable material.

The pollen which affects the hay-fever sufferers in London has probably only travelled over distances up to 80 km, but long-distance transfer is not unusual. Ships in the Atlantic have caught pollen released in North America and Tyldesley (1973) has analysed the pollen catches in the Shetland Islands to show the importance of meteorological conditions in relation to pollen movement.

CROPS DEVELOPMENT AND GROWTH

Crops and even varieties of the same crop, vary so much in their response to weather conditions, that it is not safe to make generalized conclusions. Nevertheless, it is true to say that each crop should be investigated in regard to four aspects of the problem, namely development, growth, yield and quality.

The word "development" refers to the phenological crop stages through which the crop has to pass to complete its life cycle. "Growth", on the other hand implies the production of biological matter within all the relevant stages of development. The "yield" of a crop, may be almost the total growth of crop, such as grass, although the roots are part of the growth and are not part of the yield. In a root crop the reverse is generally true, but in a crop such as sugarbeet, the roots are used for sugar production and the tops for animal fodder. In a cereal, only part of the biological matter, the seed, is usually referred to as "yield" and the same is true of fruit crops. The "quality" of a crop is a function of its use and the purpose for which it is grown.

Crop development

The development stage from sowing to emergence is best considered under the heading of germination, in which soil temperature and soil moisture are the important weather factors. In subsequent stages, temperature appears to play a dominant part, but radiation must at times be considered because the plant is not necessarily reacting to the air temperature recorded in the conventional standard way but to the temperature which is experienced by its own particular sensor and this may be dependent on the radiation balance.

In some crops, appropriate day-length is a necessary factor, irrespective of temperature or radiation. Crops have been divided into three classes, long-day, short-day and neutral (or independent of day length). Experiments designed to specify the photoperiodicity effect are simple to design, although sometimes difficult to interpret. Care must be taken to ensure uniform conditions of temperature, radiation and soil moisture. In growth chambers in particular, uniformity of experimental conditions is often more apparent than real.

Knowledge of development stages and especially ability to forecast their duration, can be of more than academic interest. In grasses, for example, the heading stage is an indicator of a change in the digestibility of the crop, an important quality in regard to both grazing and conservation as hay or silage.

Some early work by Cooper (1952) at the Plant Breeding Station, Aberystwyth, showed that while floral initiation in S.24 perennial ryegrass was a function of March day-length, ear emergence in May was linked to the accumulated day-degrees above 42°F (5.6°C) during March and April (AT), the regression being:

Date in May = $30.23 - 42.4 \log (AT/100)$

(If temperatures are measured in °C the last term becomes $-42.4 \log (AT^1/55)$.

Subsequent investigations at the Grassland Research Station, Hurley on S.24, showed that use of soil temperatures gave better results than air temperatures. Miss M. Roy found an equation which fitted Aberystwyth and Cambridge data:

Number of days to heading after April 30 = $52.6 - 2.73X$

where X was the mean earth temperature (°C) at 30 cm for March and April (Roy, 1972). As this equation does not take into account May conditions up to the actual date of heading, it has in itself some prediction value, but it was also found that general forecast guidance could be obtained from March data only, as shown in Table VIII for Aberystwyth (A) and Cambridge (C).

The ear emergence of barley also depends on the weather conditions prevailing after the critical photoperiod has been reached. Examining 10 year records of the phenology of Kenia barley grown in Hampshire Hough (1972) found that disease incidence had to be considered in addition to air temperatures. Using a simple disease scale, (1 = no mildew; 2 = slight attack; 3 =

TABLE VIII

March soil temperatures and heading dates for S.24 ryegrass

| Mean earth temp. March | Number of occasions of heading dates: | | | | | |
| | before 10 May | | 10—15 May | | after 15 May | |
	A*.	C.	A.	C.	A.	C.
Over 6.4°C	7	3	3	1	0	00
5.1— 6.4°C	0	1	3	3	2	2
Under 5.1°C	0	0	0	2	3	4

*A. = Aberystwyth, C. = Cambridge

severe attack) he found an equation for ear emergence:

Date in June = $56 - 4.9T + 7.4M$

where T is mean air temperature (°C) in May and M is the disease factor.

Hough furthermore showed that the time between sowing and ear emergence (X) had a negative correlation on the time between ear emergence and maturity. He also examined the apparent influence of many weather factors in the late July period, including rainfall total, rain days, potential transpiration and temperature. The best fits were obtained using X and rain days (over 0.2 mm), July 21—31; equations including potential transpiration were slightly less accurate whilst those using temperature and rainfall were the worst of all. Examining data from trials in several European countries carried out by the European Brewery Convention, Hough also found a good correlation (0.75) between rate of ripening and mean daily potential transpiration.

From a commercial point of view, a forecast of date of harvest can have a great economic importance. In many examples, temperature criteria have been used, more for the relative availability of information than for their inherent accuracy. There are several cases where use of sunshine, radiation or potential transpiration data has improved that standard of the forecast.

Crop growth

The start and end of crop growth may be controlled by temperature, but the growth itself is a function of energy and therefore radiation input must be taken into account. It is only in special cases, such as maize, that growth rates are strongly related to temperature and it is only in certain types of continental climate that temperature and radiation are highly correlated, so that use of temperature data implies use of radiation as a factor.

Growth can be slowed down or stopped by absence of adequate soil moisture. Any summation of an energy factor must therefore be regulated by moisture status. A convenient method of doing this was found through the concept of "effective transpiration".

TABLE IX

Estimation of effective transpiration

	March	April	May	June 1—15
Potential transpiration	1.15	1.75	3.30	2.10
Rainfall	2.85	1.40	0.75	0.20
Soil moisture deficit (max. 2 inches)	nil	0.35	2.00	2.00
Soil moisture reserve	2.00	1.65	nil	nil
Effective transpiration				
(a) from rain-fed soil	1.15	1.40	0.75	0.20
(b) from soil reserve	nil	0.35	1.65	nil
Total	1.15 +	1.75 +	2.40 +	0.20
				= 5.50 inches

Potential transpiration is itself essentially an energy factor, its use, together with rainfall, provides an estimate of soil moisture deficit*. The actual transpiration, with increasing deficit, falls below the potential level. The effective transpiration is defined as the accumulated potential transpiration over the period when soil deficits do not appreciably limit crop growth. For grass on good land the limiting deficit was found to be 50 mm (2 inches); on shallower soils this limit decreased to a value of 25 mm on the thin upland rough grazing areas. It was calculated, on a monthly basis, as in Table IX, which shows that the total effective transpiration was 5.5 inches for March 1 to June 15.

Records of rainfall and potential transpiration for 80 stations over 18 years were used in this way to find the effective transpiration for the period between start of grass growth and hay harvest. Individual values were weighted with respect to areas of grass and compared with the reported yields of hay in England, available annually on a county, regional and national basis. Provided that an annual improvement factor in the post-war years was included, which was subsequently found to be largely due to increased use of nitrogen fertilizer, a close relationship was found. The results are shown in Table X and in Fig. 3.

This and other applications of the same concept to milk production, suggests that rainfall and potential transpiration, intelligently used with due consideration of checks due to partial drought, can give good indications of the production of green matter and so act as a kind of photosynthetic measure.

---

* The amount of water needed to bring the soil back to field capacity.

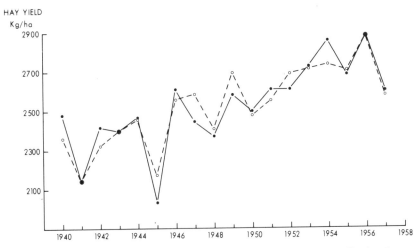

Fig. 3. Actual (full line) and estimated (dotted line) hay yields in England.

Considering grain crops, the vegetative stage and production of leaves and tillers is only part of the yield-producing process, even so, 15 years records of barley yields in southwest England show a close relationship to the potential transpiration from April to July, soil moisture rarely being a limiting factor in that area; other areas gave less promising results.

Examination of the data from the European trials referred to previously, found that barley yield seemed to depend on the following: (1) effective transpiration (limiting deficit 100 mm), braiding to ear emergence; (2) mean potential transpiration per day, ear emergence to dead ripeness, with an optimum value of 3.3 mm per day; (3) a lodging factor; (4) a disease factor.

The existence of an optimum warmth during maturation is interesting and of potential agro-climatological importance.

TABLE X

Formulae for meadow hay yields[*]

Area		Multiple correlation
North	$9.17 + 1.30T + 0.22N$	0.88
Midlands	$7.28 + 1.45T + 0.27N$	0.94
East	$2.38 + 2.22T + 0.37N$	0.94
South	$7.65 + 1.57T + 0.25N$	0.90
England	$7.05 + 1.56T + 0.27N$	0.95

[*] Where T = effective transpiration in inches; N = number of years after 1946. The yield is given in hundredweights per acre, which is approximately 125 kg/ha.

Hough's (1972) analysis of the same data indicated that a short period of high temperature about 20 days after ear emergence can reduce grain weights. High weights per 1,000 grains appeared to be associated with dry weather around ear emergence and with little or no mildew, and conversely. This indicates that weather conditions over short periods may have considerable influence on final yields. Crude analyses with monthly data can therefore only hope for moderate success.

Nevertheless, if figures for national yields are used, the importance of radiation warmth in the closing stages can be confirmed. Taking county figures of potential transpiration and weighting with respect to barley acreage to obtain a national figure, it was found for the years 1950—59 for England that there were partial correlations of 0.89 on a yearly improvement factor and 0.76 on the July potential transpiration. A smaller negative partial correlation (—0.55) was found on the June potential transpiration, which is more difficult to explain, although it is confirmed by work in The Netherlands which suggested that cool weather in June was an advantage.

If the simplest formula:

$$\text{Yield in cwt/acre} = 0.61n + 2.67J + 9.6 \text{ (multiple correlation 0.91)}$$

where J = July potential transpiration, is extrapolated for the following years results are obtained as shown in Table XI. The 1962 error is far too large, until it is known that the method of estimating national yields changed between 1961 and 1962. This emphasizes the importance of being certain that published figures are homogeneous.

The same state of affairs is shown up in the wheat yields, where again 1962 official yield figures were far higher than could have been explained by reference to the weather, using the experience of previous years.

An equation for national wheat yield based on a similar improvement factor and the national mean potential transpiration in July plus half that in August, also gave an excellent fit with a multiple regression of 0.965 over the 1950—59 period.

TABLE XI

Independent check of formula for barley yield

Year	Actual yield	Estimated yield
1960	25.0	24.5
1961	26.1	25.6
1962	29.0	25.3
1963	28.3	27.5
1964	29.4	28.4
1965	30.0	26.8 cwt/acre

Yields in the following decade, besides being influenced by new methods of assessment, had variations due to the increased incidence of diseases, chiefly mildew and rusts so that although the direct weather factors could still be recognized, they were not so dominant in determining the yields.

Many yield experiments involve variations in sowing data. In most cases the effect of date of sowing is not dependent on conditions at that time, but on the effect later in the season, when certain crop stages have to coincide in time with the most likely chances of favourable weather appropriate to those later stages.

The following example shows how the data of sowing of winter wheat affected subsequent yield and how an optimum data can be deduced from soil temperature observations.

An experiment had been carried out on Capella wheat, at up to 5 farms over 4 years, which involved 2 sowing dates and certain nitrogen treatments. By combining the fertilizer treatments, it was possible to extract 15 pairs of data, of which on 12 occasions the earlier sowing date gave the higher yield and on 3 occasions the later sowing date was more successful.

Experimental evidence suggests that wheat will grow if the temperature around seed depth is above $40°F$ ($4.4°C$), but that little growth takes place after November 30. Soil temperatures (at 09h00 GMT) were available at 10 cm depth for each of the experimental sites.

Some indication of the autumn growth will be obtained by counting the number of days between sowing date and December 1 which have such a soil temperature above the threshold ($40°F$; $4.4°C$), giving the results shown in Table XII.

These results, illustrated in Fig. 4 show that all the lower yields had growing days less than 30 or more than 60. If accumulated day-degrees of soil temperature were calculated (above $40°F$), they showed that the sum was less than 250 or more than 500. The hypothesis can therefore be put forward that as a requirement for highest yields, the autumn growth must lie within a specified range, neither too little nor too much.

This deduction is acceptable biologically but is of limited use in practical circumstances. If autumn and early winter soil temperatures remain warm, the minimum requirement of 30 days of growth before December 1 implies that October 31 is the latest advisable sowing date. So far so good, but the position is hopeless if soil temperatures fall below the threshold before the end of November. A farmer cannot be told to sow at least 30 days before an unknown date, any more than a railway traveller can be told to get off three stations before the end of the journey.

Some fixed starting date must be chosen from which the optimum range of sowing dates can be calculated. This date was selected as the first day when the 10 cm soil temperature fell below $55°F$ ($12.7°$), and the sowing day quoted in terms of the number of days after this critical date. The results are shown in Table XIII and Fig. 5.

TABLE XII

Length of autumn growth and wheat yields

Year and site		Number of days above threshold soil temperature after sowing	
		higher yield	lower yield
1957	Farm A	43	65
	Farm B	51	19
	Farm C	37	4
	Farm D	62	29
	Farm E	31	0
1958	Farm B	38	15
	Farm C	28	1
	Farm D	45	75
	Farm E	38	0
1959	Farm B	40	14
	Farm C	31	4
	Farm D	49	21
	Farm E	23	0
1960	Farm D	38	65
	Farm E	15	0

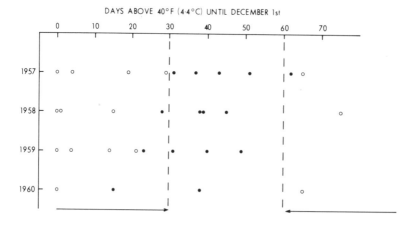

Fig. 4. Autumn growing period of wheat and subsequent yields.

TABLE XIII

Soil temperatures and sowing dates for wheat

Year and site		Number of days of soil temperatures below 55° F before sowing		Gain in yield	
		higher yield	lower yield	cwt/acre	kg/ha
1957	Farm A	28	6	0.2	25
	Farm B	29	61	1.8	225
	Farm C	45	82	1.3	165
	Farm D	16	49	3.1	390
	Farm E	47	90	2.6	325
1958	Farm B	28	52	3.6	450
	Farm C	37	74	2.7	340
	Farm D	21	—9	2.0	250
	Farm E	44	131	4.7	590
1959	Farm B	8	34	2.6	325
	Farm C	31	70	9.0	1130
	Farm D	—12	16	0.4	50
	Farm E	40	78	2.3	290
1960	Farm D	26	—1	3.1	390
	Farm E	35	122	6.5	815

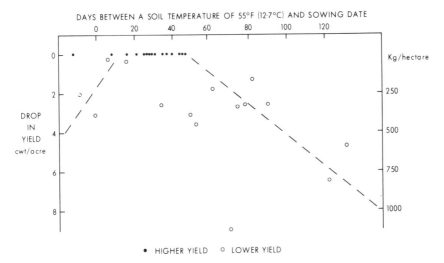

Fig. 5. Autumn soil temperature and wheat yields.

The illustration in Fig. 5 shows that the higher yields are grouped together in the time range of 10—45 days after the chosen critical date. Yields from crops sown later than 45 days after show a progressive loss of the order of 1/2 cwt/acre per week or about 10 kg/ha per day.

The practical advice to be deduced from this analysis is that this variety of winter wheat should be sown not earlier than 10 days after and not later than 45 days after the day on which the 10-cm soil temperature first falls below 55°F (12.7°C). This rule of thumb has been put into practice and has proved to be very successful. It does not take into account soil moisture conditions, but these are rarely limiting at this time of year.

Crop quality

As has been pointed out previously, the quality of grass as a fodder crop for animals is greatly dependent on the growth stage and hay cut late in the season can be of far less nutritive value than hay cut before ear emergence. A more potent weather factor, however, is that which affects the curing of the hay.

In grain crops, little work has been done on the effect of weather on its feeding quality, but the malting quality of barley is of great importance to brewers and they do not like to use a grain with a high nitrogen content.

Examination of the European trial results showed the importance of the following factors: (1) amount of nitrogen applied to the crop; (2) difference between rainfall and potential transpiration between ear emergence and dead ripeness; (3) the mean daily potential transpiration during the same period, with a critical value of 3 mm per day, at which the most nitrogen was found in the grain.

These factors are logical, because they involve the amount of nitrogen available, the ease with which the plant can take it up and the conditions during any translocation within the plant in the ripening process. It will be noted that good quality (low nitrogen) is incompatible with maximum growth, as both seem to have similar critical values of the daily warmth.

The trials in England, France, Switzerland, Ireland, Denmark, Finland, Italy, Sweden and Germany all fitted this pattern. The Netherlands were an exception, possibly because the barley there is grown over a relatively high water-table, so that the plant could use water from lower levels without drawing in nitrogen from moisture in the upper soil containing the nitrogen.

Hops are another crop in which quality is very important, the maximum possible content of alpha-acid being required for good brewing. Results of trials in six European countries on four varieties of hop show that there is a good indication of an optimum mean temperature between the time of coming into hop and date of picking. Above this temperature there is a slow decrease in quality; below the optimum the decrease is more rapid.

There is, however, some indication from recent trials that sunshine has some effect, especially in the above optimum temperature range. This is possibly because the actual temperature of the hop is different from the air temperature by an amount which depends on incoming radiation.

The optimum temperature appears to be between 16° and 17°C for the varieties Fuggle, Northern Brewer, Hallertau and Saaz, and the highest alpha-acids have been obtained, generally in Czechoslovakia, when the hours of bright sunshine over the critical period were less than 5 per day.

The one discouraging effect of such results is the difficulty of thinking how they can be put to good use, although shading might be economic in the warmer climates. If, as has been suggested, the temperature over the last two days before picking is important, then some form of hop garden heating might be envisaged, but the evidence for this weather effect is not yet conclusive.

CROP PESTS AND DISEASES

Pests and pathogens often travel by air, mostly over short distances, but occasionally on very long flights. It is therefore important to understand the meteorological influences within the realm of aero-biology. The weather is significant at all stages. To begin with, the physical conditions of the imme- diate past determine whether the pathogen (or pest) is in a condition to become air-borne, whether incubation or reproductive phase has been com- pleted. The process of release depends on the weather; for example, a change in temperature or in relative humidity, an increase of wind or convection, an impact of a raindrop can bring about a situation favourable for "take-off".

Once air-borne, only the larger insects have powers of directed flight although they may have enough energy to remain airborne and counter the feebler down currents. Otherwise, the weather conditions determine the height to which the air-borne matter is carried and the speed and direction of their subsequent movement. Furthermore, their continued viability and existence as a potential source of infection depends on such factors as the temperature, humidity and radiation. The point or area of landing, in other words the geographical destination, depends largely on the weather, although gravity obviously plays a part; larger insects, such as locusts probably con- tinue flying when over the sea, but will sink to earth over land, otherwise descents from transit height are mainly involuntary and are due to des- cending currents of air or to interception and deposition by rain or snow.

Once deposited, the weather determines the possibility of subsequent in- fection, always supposing that a suitable host or suceptible plant is available. After this stage, the whole process repeats itself, given the necessary coinci- dence in time and place, of biological processes and weather conditions.

Some of the earliest work on the long-distance transport of fungi was done by Hogg (1961) in relation to cereal rusts. The international and trans- oceanic movement of several insects was also investigated by Hurst (1968), and both these types of research involved the use of synoptic weather charts and the analysis of surface and upper air winds. Trajectories were worked out by back-tracking through a series of such charts, such a process inevitably involving an increasing radius of uncertainty the further one got back in time and distance from a known point of arrival. The reverse process, starting at a suspected source and proceeding forward in time similarly introduces a series of increasing areas, but experience has shown that such use of readily available meteorological working data can produce important and useful epidemiological conclusions.

Many pathogens, however, travel relatively short distances, possibly only a few metres or less. In such circumstances it is important to identify the critical weather influences within the release, transport and infection process. If a disease (or pest) is endemic in an area and occurs regularly each year with about the same intensity, the weather is likely to be less critical, but if the incidence and extent varies, then it is probable that there is a significant weather factor.

The most common weather-sensitive disease stage is that of plant infection and several examples could be quoted to show that knowledge of the necessary weather conditions, what they are and when they occur, is sufficient as a basis for plant disease forecasts. Such diseases are not merely of local economic importance, but can bring disaster to whole communities and the potato famine in Ireland in the middle of the 19th century is a classic example.

This was due to a combination of adverse circumstances — the over-dependence of an area on a single food crop, methods of husbandry which facilitated the overwintering of the potato blight (*Phytophthora infestans*) fungus and a succession of suitable mild, humid conditions for rapid infection. Potato blight in Europe is associated with warm or mild weather, because the climate is centred around the lower temperature infection limit of 10°C. In America, it is thought of as a cold weather disease, because this lower threshold is always cleared and the disease is more likely to be limited by high temperatures (and low humidity).

Field investigations by Beaumont (1947) in the 1920's showed that 48 h of weather with temperature over 10°C and relative humidities over 75% were necessary infection conditions. These "Beaumont periods" are still used as a basis for disease warnings in the United Kingdom, making use of standard weather observations at synoptic stations. Subsequent work showed that periods of relative humidity over 90% could also be successfully used, but were less convenient. In Ireland, such weather can be identified from observations of warm sectors on a synoptic map and other criteria are used in The Netherlands and Germany, but basically all such criteria are aimed at defining periods of warm humid weather.

The duration of wetness of a plant leaf is often a critical factor in determining the possibility of infection. In the case of apple scab (*Venturia in-*

aequalis) the precise requirements were established in the laboratory by Mills (1954) giving rise to the "Mills periods". Apple scab is a disease occurring first in the spring, when asco-spores which have over-wintered in the orchard top soil are released by the effect of rain. Mills' work showed that if such spores land on an apple leaf, then infection will occur if the leaves are wet for a relatively short period (about 10 h) in high spring temperatures, but at lower temperatures near freezing point, as long a wetness period as 48 h may be necessary.

It was first decided to verify these criteria in the field under English conditions, but to do this a means of measuring surface wetness had to be designed. This was done by constructing a delicate type of dew balance, which, by weighing showed on a time chart the number of hours an expanded polystyrene block retained a moisture cover. This instrument was accurate enough to record dew fall and indicate the periods of wetting, the duration of full wetness and the period of drying.

Exposure of such wetness recorders in an orchard at the height of the lower leaves of the apple tree, together with thermographs (in a Stevenson screen) proved that the Mills criteria were valid and could be used for disease prediction when wetness recorders were installed.

The next step was to examine the (standard) meteorological conditions under which the wetness persisted. The nearest reporting station to the orchard was 30 km away, but even so a reliable link was found between the wetness duration and the length of time the relative humidity in the screen remained at 90% or above following rainfall. The main discrepancies occurred when the periods of wetness were small and insignificant with respect to disease infection.

It was now possible to use a network of synoptic stations (taking hourly readings) for identifying the significant weather. Mills periods, depending on surface wetness, were thus replaced by Smith periods depending on screen humidities. Furthermore, past weather could be analysed from such standard records and their common features could be studied.

For example, if the disease-significant occasions are considered when there was less than 12 h continuous rain, it was found that:

If it begins to rain at 06h00 GMT it has to rain for 12 hours;
if it begins to rain at 10h00 GMT it has to rain for 8 hours;
if it begins to rain at 14h00 GMT it has to rain for 4 hours;
if it begins to rain at 16h00 GMT it has to rain for 2 hours.

and in all cases the rain has still to be falling at 17h00 GMT. From this and similar considerations it was possible to formulate a set of rules which a fruit-grower, using only a thermometer and making two inspections of his orchard each day, could use to identify the infection periods. They were as follows:

Infection is likely if: (a) rain starts to fall during the day, is still falling at 17h00 GMT and the temperature at that time is 50°F or above; (b) rain starts to fall during the evening or night and the leaves are still wet at 08h00 GMT with a temperature at that time of 50° F or above; (c) rain is falling or leaves are wet following rain on any two successive inspections at 08h00 GMT and 17h00 GMT with the mean of the temperature at these times being 44° F or above; (d) similarly, for three successive inspections and a mean of 42° F or above; (e) similarly for four successive inspections and a mean of 40° F or above; (f) rain is falling or leaves are wet following rain on any five successive inspections, irrespective of temperature.

Tests on 242 separate occasions showed that 102 satisfied these criteria and of these 94 were correct with 8 false alarms; the remaining 140 "negative" occasions giving only 1 occasion when disease would have been likely.

This case study of apple scab is an excellent example of how good research in a laboratory can, by successive steps, lead to the establishment of reliable disease criteria in the field, using either delicate instruments, existing weather observing networks or rules for single-observer decisions.

In attempts to forecast the incidence of potato blight it is often assumed that in all cases sufficient inoculum will be present. In the case of apple scab, disease infection criteria are not applied until it is known that the spring release of asco-spores has started and the rain which occasions their release is also a cause of the necessary leaf wetness. In other diseases, such as barley mildew (*Erysiphe graminis*), the infection conditions seem to be of a secondary importance and the spread of the disease seems to depend largely on the spore concentration on any particular day.

This was shown by experiments at the Plant Pathology Laboratory at Harpenden, where a succession of barley plants were exposed on any possible exterior infection for only 48 h at a time, the number of disease pustules being counted after incubation under conditions of quarantine, so that it was known when the infection occurred. The plants showing the highest pustule counts were those exposed during the days with high spore counts (observed continuously in an aspirated spore-trap.)

Disease prediction therefore depends on the identification of the conditions which produce high spore counts. This will depend on the previous history of the pathogen (and on the previous weather) because the count builds up in May and June with each successive generation of spores. This increase starts earlier in winter-sown barley than in a spring-sown crop. Observations have shown that spore concentrations are greatly increased in May if the wind is blowing from a winter barley field to a spring barley site.

Where no such complications occur, the spore-trap counts can be analysed with respect to the daily weather. Firstly, a count has to be taken of the spore counts which occurred on the days earlier in time by one, two or more incubation periods. This, in effect, gives a measure of the spore release potential. Mildew spores are released chiefly by the wind-caused movement of the barley leaves, provided that they have experienced appropriate conditions of temperature and moisture.

Analysis of the daily spore counts shows that they increase with increase in maximum temperature, wind strength and sunshine, with indications of rapid increases if the day maximum exceeds 17°C and the run-of-wind exceeds 340 km per day.

To institute a useful disease warning service it was needed to know when it would be most effective to put on a single spray treatment. Spray-timing trials at several centres over some three years indicated that this was likely to be when the spore count first showed a major increase of (one or even two) orders of magnitude. This appeared to occur when there had been a succession of days, each separated by an incubation period, when warm, sunny, windy weather was suitable for spore release and dispersal.

The next problem was the best way in which to use data from synoptic reporting stations to identify such critcal days and the following formula was chosen: T = day maximum temperature (°C); W = 12h00 GMT wind speed (knots); H = hours of sunshine; (the effect of rain was ignored although it was known that prolonged rain was associated with low spore counts, but in that case both T and H would be low in value).

The weather index was taken to be:

$3T + 1/2W + H$

and the critical value of this index was 65, at and above which conditions were deemed suitable for a high spore count.

The best spray data was then chosen as the second of such high index days after mildew had first been observed in the crop, provided that such days were separated by not less than 5 and not more than 9 days. In other words, two successive generation spore build-ups were being sought from the initial infection. Tests at various experimental sites show that this (and similar) criteria have high hopes of success, and if so, yet another disease can be combatted on a sound scientific basis with the maximum gains in yield from the least expenditure in fungicides.

Some diseases are transmitted by insect vectors, for example the virus yellow disease of sugarbeet is vectored by aphids. An assessment of the weather affecting the spring increase in the aphid population can therefore be used as an estimate of future disease severity. If the weather affecting the aphids can be identified then disease forecasts can be made. The significant conditions in this case seem to be the severity of the winter (hence the size of the overwintering population) and the warmth of late winter and spring (which controls the speed of population increase).

A search through the published papers in plant pathology and entomology will provide many other examples of weather-sensitive pests and diseases. Success in discovering the significant factors depends largely on the availability of data and the skill of the analysis which should preferably be carried out by the collaboration of agricultural meteorologists with the agricultural scientist concerned.

The unknown pathogen

A fatal human disease of unknown origin occurs in Yugoslavia, Bulgaria and Rumania, chiefly in the valleys of the Danube and its tributaries. Despite intensive medical research over many years, no cause can be found for this disease, known as endemic nephropathy. All the more likely sources of infection or contamination have been checked without result. The only disease with similar symptoms is found in pigs in Denmark and is known to be due to a mycotoxin, but this fungus is not found in the Balkans.

The disease appears to be confined to certain villages, and possibly to certain houses or families in such villages. It is not, however, hereditary, because anyone leaving such a family to live elsewhere at an early age is unlikely to die from the disease. On the other hand, a daughter-in-law marrying into the affected family will contract the disease and yet there is no evidence that it is venereal. There is a tendency for women to be more susceptible than men, and because most of the food eaten by the villagers is home-produced and almost certainly stored within the home in loft or cellar, plus the fact that a high proportion of the affected families have their own food-producing garden or allotment, then it is reasonable to suggest that another mycotoxin on food is involved. Such a pathogen must clearly be incapable of easy spread and have possibly severely restricted environmental requirements.

If, however, this unknown fungal growth is a factor, then there should be some relationship between year-to-year deaths and the preceding weather, because the growth, harvest and storage physical conditions will vary with successive seasons. Most fungi flourish in conditions of high relative humidity, but this element cannot be satisfactorily extracted from standard meteorological records. There is, however, one way of specifying the weather which is likely to have long periods of high relative humidities and that is by using the positive difference between rainfall and evaporation on a monthly basis. The significant months are likely to be those of harvest and post-harvest, namely August to November, for humidity conditions during the subsequent winter are unlikely to vary much from year to year. If it is impossible to estimate the evaporation factor, some use could still be made of rainfall totals alone.

If a mycotoxin is involved, the effect must be cumulative and before any estimate is made of the effect of the final large dose leading to a death, it is thus preferable to know the morbidity figures to get some indication of those at maximum risk. Yearly mortality and morbidity figures were available for the district of Brodska Posavine in Yugoslavia, and it was possible to obtain local montly rainfall data and an estimate of evaporation.

To serve as an illness factor, it was decided to add together the numbers reported as suffering from the disease over three successive years (e.g. years 1, 2 and 3) and to subtract the number reported as dying in the last two years of this triplet (e.g. years 2 and 3). This should give a good idea of the number of people at maximum risk in the next year (year 4). By hypothesis these should be influenced by the weather of the previous year (year 3).

The chosen weather parameter was the sum of: the excess August rainfall over 80 mm; plus the excess September rainfall over 60 mm; plus the excess October rainfall over 30 mm; plus the excess November rainfall over 20 mm.

During the 10 years 1960—1969, the partial correlations were as follows: deaths on illness factor (I) + 0.81; deaths on wheather factor (W) + 0.77 with a formula

Deaths = 0.05 W + 0.23 I − 1.4

the data being shown in Table XIV.

Thus the weather of year 3 appears to have an appreciable effect on the deaths in year 4, when account is taken of the number of people likely to be at maximum risk. It is interesting to note that the weather of year 2 appears to have no significant effect.

The hypothesis of a weather-sensitive mycotoxin has therefore been justified by the Yugoslav analysis, but for reinforcement the same type of relationship should hold elsewhere. Attention was therefore turned to the district of Erghevitza in Rumania, using rainfall records from Turnu-Severin. Morbidity data were not available and it was not possible to make a satisfactory estimate of evaporation so that the weather factor was taken as the September to November rainfall. August was omitted because in none of the years under review was August a wet month (this was not the case in Yugoslavia where there were two very wet Augusts).

Because no illness factor could be formulated, it was decided to take the weather factor in pairs of successive years, on the theory that the weather

TABLE XIV

Verification of estimates of deaths from nephritis in Yugoslavia *

Year	W	I	Estimated deaths	Actual deaths
1960	132	24	11	9
1961	200	26	15	15
1962	77	48	13	12
1963	108	49	15	16
1964	115	55	17	17
1965	142	38	14	14
1966	176	37	16	12
1967	87	30	10	10
1968	98	27	10	11
1969	193	38	17	19
1970	120	31	12	15

* For explanation of symbols see text.

TABLE XV

Verification of estimates of deaths from nephritis in Rumania

Death years	R	R_p	Estimated deaths	Actual deaths
1958—60	161	137	18.0	21.5
1959—61	254	161	28.6	28
1960—62	193	254	32.5	35
1961—63	212	193	28.1	30.5
1962—64	121	212	22.0	22
1963—65	148	121	15.3	17.5
1964—66	223	148	24.6	19.6
1965—67	89	223	20.3	17
1966—68	222	89	18.6	18

*For explanation of symbols see text.

factor of the first of the pair would affect the illnesses occurring in the second of the pair.

The weather factor was therefore taken to be: rainfall during the 3 months, Sept., Oct., Nov. in year 1 (R_p); and rainfall during the 3 months, Sept., Oct., Nov. in year 2 (R).

The deaths, the dependent variable, in a second attempt to obviate the absence of morbidity data, were summarized as: D = deaths in year 2 plus deaths in year 3 plus 1/2 deaths in year 4.

Partial correlations were found as follows: D and R + 0.84; D and R_p + 0.87; and a regression equation: deaths = $0.088R + 0.1R_p - 9.85$.

No significant effect could be traced for the rainfall in the year previous to R_p. (Details are given in Table XV.)

The multiple correlation of weather and deaths is 0.89 which again supports the hypothesis that some weather-sensitive pathogen could have an influence on mortality from this disease.

The third country to be studied was Bulgaria and monthly mortality data were available for Vratza county; rainfall records were used from Vratza and Slatina. It was possible to make a reasonable estimate of evaporation, namely: August 75 mm; September 50 mm; October 20 mm; November 10 mm.

Errors in such an estimate, or year-to-year variations therein, will not have a major effect as the rainfall totals are the dominant weather factor.

Again, as for Rumania, the weather factor was taken over a couple of successive years, namely: R_p is excess rain in year 1; R is excess rain in year 2.

The deaths factor was also taken over 2 years, this time without any overlap. The period October in year 2 to September in year 4 was used thus making sure that if food was responsible it was confined to produce harvested in years 1 and 2.

TABLE XVI

Verification of estimates of deaths from nephritis in Bulgaria

Years of deaths	R	R_p	Estimated deaths	Deaths
1961 Oct.—1963 Sept.	51	30	30	34
1962 Oct.—1964 Sept.	107	51	66	47
1963 Oct.—1965 Sept.	72	107	80	64
1964 Oct.—1966 Sept.	125	72	85	75
1965 Oct.—1967 Sept.	37	125	73	80
1966 Oct.—1968 Sept.	167	37	85	94
1967 Oct.—1969 Sept.	27	167	90	110
1968 Oct.—1970 Sept.	140	27	69	88
1969 Oct.—1971 Sept.	44	140	83	73

The formula was found to be:

$$D = 0.51\,R_p + 0.44\,R - 7.3$$

with a multiple correlation of 0.80 (see Table XVI).

If the deaths in the calendar years 3 and 4 were used as the death parameter a very similar result was obtained.

Thus the same type of inference could be drawn in each of the three countries. In Bulgaria, however, one further relationship could be found, namely that between the weather factor of years 1 and 2 and the deaths in year 4. This can be shown by the data in Table XVII. The two sets of data in the table have a correlation of 0.87, and if partial correlations are found for

TABLE XVII

Evidence of delayed mortality in Bulgaria

Sum of weather factors over two successive years	Death two years later
81	11
158	35
179	37
197	39
162	45
204	52
194	55
167	32
184	37

each of the two rainfall years they are 0.85 and 0.89. The simple regression is 0.3 $(R + R_p) - 13.2$. This relationship, if it is not just a mathematical accident, implies a time lag of 24 months or more while in other countries the time lag appeared to be more like 12—18 months.

It is therefore interesting to find that patients in Bulgaria are known to be able to withstand the terminal stages of the disease longer than others elsewhere. In Rumania, in particular, the disease is often only diagnosed when the sufferers are admitted to hospital for other reasons.

The pathogen is still unidentified, but there now seems ample reason to believe in the existence of a mycotoxin. Considering that all other attempts to find a source for this disease have failed, this discovery illustrates the high potential value of an agro-meteorological investigation. The results will not finalize the problem, but at least research efforts can now be concentrated on the potentially profitable avenues of investigation.

In general it may be said that almost all human diseases caused or vectored by biological activity have a meteorological factor and that the surface has hardly been scratched in problems of medical meteorology.

It must not be thought that the simple form of the results quoted in relation to endemic nephropathy imply that they are reached without difficulty. They are the result of trying out every sensible combination of the restricted available data. They are presented here because they provide the most convincing evidence justifying the existence of a so far unidentified mycotoxin.

However clever is the mathematical manipulation of the figures, it is the pathological and medical logic implied in the final result which can be accepted as an approach to the truth. The degree of correlation tells nothing except the degree of risk of a pure chance agreement.

The meteorological parameter is not the ideal one, but the best use has to be made of the available data. The medical statistics may be subject to variable standards of diagnosis or record. Nevertheless, if a similar type of weather—death relationship is found using three distinct sets of data, and in each case the correlation is highly significant and unlikely to be encountered by pure chance, then only two conclusions can be drawn. Either the weather during and after harvest does have an effect on something which causes subsequent death, or else there is a third factor which is affecting both the autumn rainfall and the deaths during the following years. The second alternative is almost impossible to conceive. Therefore the only conclusion that can be drawn is that weather does affect the unknown pathogen, unless we have encountered a series of highly improbable arithmetical coincidences.

REFERENCES

Beaumont, A., 1947. The dependence on the weather of the dates of outbreak of potatoe blight epidemics. *Trans. Br. Mycol. Soc.*, 31: 45—53.

Cooper, J.P., 1952. Studies on growth and development in Lolium, 3. Influence of season and latitude on ear emergence. *J. Ecol.*, 40: 352—379.

Mills, J.D. and La Plante, A.A., 1954. Diseases and insects in the orchard. *Cornell Extr. Bull.*, 711: 20—22.

SUGGESTIONS FOR FURTHER READING

Pollination

Hurst, G.W., 1967. Honey production and summer temperatures. *Meteorol. Mag., Lond.*, 96: 116—120.

Hurst, G.W., 1970. Temperatures in high summer and honey production. *Meteorol. Mag., Lond.*, 99: 75—82.

Davies, R.R. and Smith, L.P., 1973. Forecasting the start and severity of the hay fever season. *Clin. Allerg.*, 3: 263—267.

Davies, R.R. and Smith, L.P., 1973. Weather and the grass pollen content of the air. *Clin. Allerg.*, 4: 95—108.

Tyldesley, J.B., 1973. Long-range transmission of tree pollen to Shetland, I. Sampling and trajectories; II. Calculation of pollen deposition; III. Frequencies over the past hundred years. *New Phytol. Cambr.*, 72: 175—181; 183—190; 691—697.

Crop development and growth

Smith, L.P., 1960. The relation between weather and meadow hay yields in England, 1939—56. *J. Brit. Grassl. Soc.*, 15: 203—208.

Smith, L.P., 1962. Meadow hay yields. *Outlook on Agric.*, 3: 219—224.

Croxall, H.E. and Smith, L.P., 1965. Sowing dates for winter wheat. *NAAS Q. Rev.*, 68: 147—149.

Smith, L.P., 1967. The effect of weather on the growth of barley. *Field Crop Abstr.*, 20 (4): 273—278.

Smith, L.P., 1968. Effective transpiration; a meteorological parameter for grassland. *Unesco, Nat. Resour. Res.*, 5: 429—433.

Smith, L.P., 1969. The effect of weather on the quality of hops. *Ann. Rep. Dept. Hop Res., Wye Coll.*, 1969: 47—60 (also contained in the Unesco publication, *Ecology and Conservation*, 5: 99—106).

Roy, M.G. and Peacock, J.M., 1971. Seasonal forecasting of the spring growth and flowering of forage crops in the British Isles. *Univ. Coll. Wales, Aberystwyth, Mem.*, 14: 129—155.

Roy, M.G., 1972. Heading dates of S 24 perennial ryegrass and spring temperatures. *J. Brit. Grassl. Soc.*, 27: 251—260.

Cochrane, J., 1972. Climate limits for viticulture in England and Wales. *J. Engl. Vineyards Assoc.*, 6 (July): 35—37.

Gloyne, R.W., 1972. Some major meteorological influences on crop production in higher latitude, maritime areas. *Biometeorology*, 5, Part 1, p. 56.

Hough, M.N., 1972. Weather factors affecting the development of maize from sowing to flowering. *J. Agric. Sci.*, 78: 325—331.

Smith, L.P., 1972. The effect of climate and size of farm on type of farming. *Agric. Meteorol.*, 9: 217—223.

Crop diseases

Smith, L.P., 1953. Forecasting outbreaks of potato blight. *Meteorol. Mag., Lond.*, 82: 113—115.

Hogg, W.H., 1956. Weather and the incidence of chocolate spot in beans. *NAAS Q. Rev.*, 32: 87—92.

Smith, L.P., 1956. Potato blight forecasting by 90% humidity criteria. *Plant Pathol.*, 5: 83—87.

Smith, L.P., 1958. The duration of surface wetness. *Proc. Int. Hort. Congr. 15th, Nice*, 3, pp. 478—484.

Hogg, W.H., 1961. The use of trajectories in Black Rust epidemiology. *Coloquis Europeo sobre la Roya Negra de los Cereales*, 1961, pp. 4—8.

Preece, T.F. and Smith, L.P., 1961. Apple scab infection weather in England and Wales, 1956—60. *Plant Pathol.*, 10 (2): 43—51.

Hogg, W.H., 1962. The use of upper air data in relation to plant diseases. *Univ. Coll. Wales, Aberystwyth, Mem.*, 5: 22—28.

Smith, L.P., 1962. Simplified weather criteria for apple scab infection. *Proc. 16th. Int. Hort. Congr., 16th, Brussels*, 3, II, pp. 226—229.

Hurst, G.W., 1965. Forecasting the severity of sugar-beet yellows. *Plant Pathol.*, 14 (2): 47—53.

Smith, L.P. and Walker, J., 1966. Simplified weather criteria for potato blight infection periods. *Plant Pathol.*, 15 (3): 113—116.

Hirst, J.M., Stedman, O.J. and Hogg, W.H., 1967. Long-distance spore transport; methods of measurement, vertical spore profiles and the detection of immigrant spores. *J. Genet. Microbiol.*, 48: 328—355.

Hurst, G.W., Hirst, J.M. and Stedman, O.J., 1967. Long-distance spore transport; vertical sections of spore clouds over the sea. *J. Genet. Microbiol.*, 48: 357—377.

Hogg, W.H., 1968. Wheat rusts, meteorological factors and warning systems. *Proc. Cereal Rust Conf., Ceiras, Portugal*, pp. 36—39.

Hogg, W.H., 1969. Meteorological factors affecting the epidemiology of wheat rusts. *W.M.O. Tech. Note*, No. 99 (see Appendix I).

Smith, L.P. and Davies, R.R., 1972. Weather conditions and spore trap catches of barley mildew. *Plant Pathol.*, 22: 1—10.

Polley, R.W. and Smith, L.P., 1973. Barley mildew forecasting. *Proc. Brit. Insecticide Fungicide Conf., 7th, Brighton*, 2, pp. 373—378.

Crop pests

Hurst, G.W., 1964. Effects of weather conditions on thrips activity. *Agric. Meteorol.*, 1: 130—141.

Hurst, G.W., 1964. Meteorological aspects of the migration to Britain of *Laphygma exigua* and certain other months on specific occasions. *Agric. Meteorol.*, 1: 271—281.

Hurst, G.W., 1965. Meteorology and locust migrations. *Nature, Lond.*, 205: 661—662.

Hurst, G.W., 1967. Take-off thresholds in thysanoptera and the forecasting of migratory flight. *Int. J. Biometeorol.*, 2 (part 2): 576—578.

Hurst, G.W., 1968. Aerial infiltration by windborn insects and spores. *Unesco, Nat. Resour. Res.*, 7: 153—156.

Hurst, G.W., 1968. Insect migration. *Proc. W.M.O. Reg. Training Sem., Wageningen, May 1968*, pp. 163—176.

Hurst, G.W., 1969. Insect migrations to the British Isles. *Q. J.R. Meteorol. Soc.*, 95: 435—439.

Shaw, M.W. and Hurst, G.W., 1969. A minor immigration of the diamond-back moth. *Agric. Meteorol.*, 6: 125—132.

CHAPTER 4

The Modes of Agricultural Meteorology — Animals

> "The world was made to be inhabited by beasts,
> but studied and contemplated by man."
>
> THOMAS BROWNE

ANIMAL REPRODUCTION

Reproduction in domesticated farm animals can hardly be called a highly efficient process, and the economic consequences of reproductive disorders are greater than those which are the consequence of other animal diseases. Failures have often been ascribed to inefficient management of stock, but it is by no means certain that this is not a convenient and over-simplified explanation covering a wealth of scientific ignorance.

Some idea of the losses involved can be obtained from the data obtained from many sources which show that for every 100 ova released from the ovaries, only about 75 are fertilized and implanted in cattle and pigs, and as few as 50 in sheep. Further losses occur; early embryonic death, between the time of implantation and the end of the first third of the duration of pregnancy, accounts for an additional 25% of reproductive failure, and abortion and stillbirth in the remaining two-thirds produces another 10%. It is also important to realize that well over half of the cases of bovine abortion have no attributable cause, in other words, "no diagnosis".

This failure to explain is characteristic of the whole complex of animal infertility and yet it is not unreasonable to state that the direct and secondary effects of weather have a bearing on this important problem and await elucidation. The environmental conditions could affect fertility and reproductive efficiency before attempted conception, during pregnancy, and at birth. Such conditions might place a direct physical stress on the animal due to excess rain or snow, heat or cold, which can be called the exposure factor. Indirectly the weather will influence the natural or conserved quantity and quality of animal fodder. In particular the growing conditions may have determined the presence or absence of fungal hormones or toxins, for the quality of animal food does not only imply an analysis in terms of calories or carbohydrates. Small amounts of pollutants, as yet undiagnosed, may be a critical factor. The effect of weather on animal nutrition, in its widest sense, is therefore a matter of great importance.

Although only a limited amount of agro-meteorological work has been done in this important subject, some examples can be quoted of attempts to identify the critical weather influences.

Twin-lambs

Farmers have long since recognized that in order to obtain high proportion of twin-lambs in a flock, it is good standard practice to allow the ewe access to good food supplies before she is put to the ram, a process known as "flushing". The importance of this rising plane of nutrition has been confirmed by experimental work in countries such as New Zealand. However desirable this practice may be, it is not always easy to carry out if grass is not growing freely in the fields in late summer, for supplementary feeding may be uneconomic.

A promising meteorological parameter for grass growth has been found in "effective transpiration", that is, the sum of the potential transpiration over the period when the soil-moisture conditions within the main grass root zone was not limiting for full growth (see Chapter 3, p. 78).

If then the effective transpiration for May and June was A, and that for July and August was B, then the relative and absolute values of A and B should provide an indicator to the process of flushing the ewes. If both A and B are high and plenty of grass is available, the difficulties are solely those of management. If A is less than B, then the weather has provided the best conditions for increasing natural food supply; if A is greater than B, due to dry weather in the latter months, then flushing is going to be difficult to accomplish.

Data on twinning were available for a large sample (about 20,000) of ewes in Denbighshire, Wales, over a period of 11 years. Areal values for A and B were found by weighting the records of rainfall and calculations of potential transpiration from a reference station to a county value, subsequently calculating the effective transpiration. The correlation between the selected weather parameter, namely $100A/B(A + B)$, which takes into account the probable effects of grass availability, and the twinning percentage was found to be -0.84.

Although plentiful rain in July and August would help to ensure a good supply of grass, an excess of rain during this period would be detrimental to the ewe, especially if the weather is cold or sunshine is deficient. A measure of this stress factor could be taken as $(R - C)/C$, where R is the July—August rainfall and C is the potential transpiration for the same period.

A double regression of twinning percentages on these two meteorological parameters can now be derived, with a multiple correlation of 0.87, suggesting that the analysis has dealt with the major weather factors. Sunshine, taken alone, was not found to produce any subsequent improvement, but its effects have already been included in the C factor (see Fig. 6A).

The exercise was repeated using more than one sample station as a reference point from which the county values were obtained. No great differences from the simple approach were found, some years providing a better fit to the observed twinning, others being not so successful.

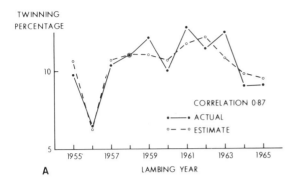

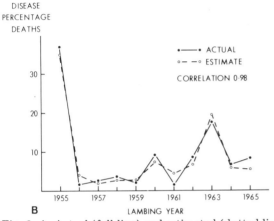

Fig. 6. A. Actual (full line) and estimated (dotted line) twin-lamb percentages.
B. Actual (full line) and estimated (dotted line) deaths from pregnancy toxaemia (twin-lamb disease).

TABLE XVIII

Verification of estimates of twin lamb percentages*

Year	Twinning percentage	$\dfrac{100\,A}{B\,(A+B)}$	$\dfrac{R-C}{C}$	Twinning estimate	Error
1955—6	6.6	27.8	0	6.3	—0.3
1963—4	9.0	15.1	0	9.8	+ 0.8
1964—5	9.1	16.0	0	9.5	+ 0.4
1954—5	9.7	10.0	0.33	10.7	+ 1.0
1959—60	10.0	11.6	0	10.7	+ 0.7
1956—7	10.4	6.6	1.14	10.7	+ 0.3
1957—8	11.1	6.7	0.73	11.2	+ 0.1
1961—2	11.4	6.2	0	12.2	+ 0.8
1958—9	12.2	8.5	0.35	11.1	—1.1
1962—3	12.4	11.4	0	10.8	—1.6
1960—1	12.8	6.9	0.28	11.7	—1.1

*For explanation of symbols see text.

The results for the simplest approach are shown in Table XVIII.

The correlations of the twinning percentages on the flushing factor and on the excess rain factor were —0.8358 and +0.2195 with a correlation between the weather factors of —0.5005. The partial correlations were therefore: twinning and 100 $A/B(A + B)$: —0.86; twinning and $(R - C)/C$: —0.42; with a multiple correlation of —0.87, and a formula:

Twinning percentage = $13.84 - 0.27\,u - 1.17\,v$

where u refers to the flushing factor and v to the excess rain factor. The results are illustrated in Fig. 6A; Fig. 6B refers to "twin lamb disease" (see p. 103).

Cow fertility

The best available data to examine in relation to the possible effects of weather on bovine infertility are the records of artificial inseminations carried out by the centres under the control of the Milk Marketing Board. There are 23 such centres, covering the greater part of England and Wales, the only exceptions being areas served by other organizations, namely the Hertfordshire, Cambridge, West Suffolk and West Norfolk area, the Hereford, Worcester, Oxford, Berkshire, Surrey and Hampshire area, Somerset and South Devon.

If the first A.I. service is unsuccessful, the customer is supplied with a second (free) service 30 days later. The success is judged by the percentage of customers who do not return for a third service 60 days after the first attempt; this is known as the non-return rate and details are available on a monthly basis for a number of years.

This non-return (or success) rate varies, on average, between about 70% cent in southeast England to over 80% in north Wales. The commonly accepted explanation for this variation is the variation in size of herd, it being thought that in a smaller herd, the dairy farmer or his cowman has a better chance of correctly identifying the time when the cow will be receptive to service. As the average herd size in Essex is about 40, and that in Anglesey only 10, this might seem to be a convincing argument. However, herd size and farm size are highly correlated with climate, in particular with the average effective transpiration over the growing season, and hence the suitability of the area for grass production. Furthermore, during 1959—60, following an exceptionally dry summer of 1959, success rates were outstandingly low at all centres. Another feature which introduces doubt is that although herd sizes have been slowly but continually increasing over the last one or two decades, there has been no evidence of a decline in success rate. Although this can be explained away by the suggestion of better management, it does seem possible that weather has a so far undefined influence.

In the case of twinning of lambs, use was made of effective transpiration because it was necessary to form an estimate of grass growth. In the present problem, the reverse side of the picture seems more promising. Contributions towards the total of effective transpiration were not made when the soil moisture deficit was too large to permit full grass growth. If the number of days when this moisture stress is found, called grass non-growing days, or more simply "dry-days", then some idea is found of the length of period when the grass quality is likely to be different from that grown under conditions of no moisture stress.

The area served by each A.I. centre was known, so that an estimate of mean herd size could be found from annual returns which give the number of cows in milk for each county and the number of registered milk producers. The mean for the A.I. centre was found from totals and not by averaging the county means. The number of dry-days were found for about 100 sample stations during the years under review (1963—1966), and the A.I. centre mean was found by weighting with respect to the number of cows in the area around each sample meteorological reporting station situated within the A.I. centre service region.

This data is not ideal, because not all farmers use the A.I. service, especially those with very large herds who are more likely to keep their own bull, but the great majority of them rely on the M.M.B. semen; also the selection of sample rainfall stations may be inadequate, but the two sets of independent data are probably of comparable accuracy. The results are shown in Table XIX. Total correlations of success rate on dry days and on herd size are —0.90 and —0.64, respectively. The equivalent partial correlations become:

success rate on dry days (constant herd size) —0.83
success rate on herd size (constant dry days) —0.20

This suggests that there is an unidentified weather effect which is more important than herd size in affecting the A.I. successes.

Attention was then turned to the monthly data, remembering that the figures quoted for any one month refer to attempted inseminations one or two months earlier. It was found that all areas had their lowest successes at the end of the winter, and their highest usually in early summer. In the eastern areas, the success fell away considerably in late summer and autumn, but the corresponding change in western areas was much less. This seasonal variation again appears to imply a reflection of nutritional conditions.

Partial correlations were calculated for each separate month (see Fig. 7). Those for herd size showed a random variation, all being below —0.50. Those for dry days showed highest values from November (—0.79) to March (—0.78) and a very low value for August (—0.27). The August result would only be expected, dealing as it does with unsuccessful A.I. services in June and July, if indeed the growth of grass is a significant factor, for the cows would have been feeding on the good growth of May and June. If dry-days

TABLE XIX

Artificial insemination centres, weather and herd size (1963—1966)

A.I. Centre	Approximate area served	Mean dry days	Mean herd size	Success rate
Llangefni	Anglesey, Caernarvon	2	11	80.1
Ruthin	Denbigh, Flint	11	15	79.5
Burley	W. Riding	16	18	79.1
Penrith	Northern England	5	22	78.5
York	N. and E. Ridings	16	26	78.3
Welshpool	Central Wales	13	19	78.1
Whalley	Lancashire	16	21	77.6
Carmarthen	S. Wales	11	15	77.4
Felin Fach	Cardigan	7	13	77.3
Buxton	Peak District	14	21	77.3
Torrington	N. Devon	13	16	76.9
Cheswardine	Salop; W. Stafford	26	24	76.2
Sturminster Newton	Dorset	38	35	75.8
Gloucester	S.W. Midlands	24	28	75.6
Tarporley	Cheshire	23	28	75.5
Tean	Stafford and S. Derby	31	22	75.5
Praze	Cornwall	29	14	75.4
Sutton Bon-ington	E. Midlands	42	22	74.6
Little Horwood	S.E. Midlands	36	25	74.5
Clyst Honiton	E. Devon	41	16	74.5
Beccles	E. Suffolk and Norfolk	41	27	73.7
Whiligh	S.E. England	34	31	73.2
Writtle	Essex	49	37	71.7

are important September would be the first month when any delayed effect would be noticed.

Attempts to link these results with hay quality, based on the weather during hay-making in June were unsuccessful, but as hay-making time and the taking of autumn grass for silage varies from area to area, this result was disappointing but not surprising.

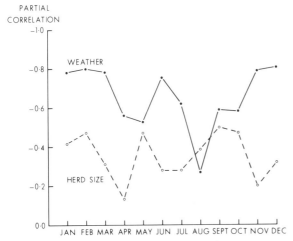

Fig. 7. Partial correlation coefficients of success rates of artificial insemination of cows on weather factor and herd size.

The explanation of any cause of infertility due to grass quality influenced by dry conditions, is probably very complex with no simple answer. For instance, it may be that grass growing freely in good soil-moisture conditions contains some ingredient, such as a hormone, which promotes fertility. On the other hand, mycotoxins may flourish in dried-out pastures or when rain returns to such pastures, producing adverse effects until grazing ceases in late autumn.

The problem remains unsolved but enough has been done to show that the weather factor cannot be ignored.

Exposure factor during pregnancy

The Denbighshire data for lamb-twins also contained details of the number of ewes who died from "twin-lamb disease", a form of pregnancy toxaemia. The incidence of such losses can vary from as little as 1% to over a third of the twin-carrying ewes in a bad season.

In was thought that the weather stress played a large part in this problem and that wind, cold and a continuous wet fleece from frequent rain or snow would debilitate the ewe.

No influence of wind strength could be found to fit the data, although it would be unwise to discard this parameter on limited evidence, as there are signs that it has an effect on incidence of these losses in the eastern Pennines. A great deal obviously depends on the degree of topographical shelter available to the flocks, as well as husbandry practice and provision of artificial shelter or incorporation of tree or wall shelter-belts.

Clear evidence was found of the influence of cold and wetness. Moreover, a combination of the two factors was more important than either component taken separately. Linear relationships over the temperature and rainfall range were unlikely, because under the milder conditions the stress of exposure is probably small to a hardy hill animal. Beyond certain thresholds, however, the pregnant ewe is faced with a dilemma with no favourable choice. Either she continues to seek her food and suffers the full force of exposure to adverse weather, or else she seeks any available shelter and abandons her search for the limited grazing. In both cases she uses up more energy than she puts back in food, her condition deteriorates, and she can no longer carry her twins.

To obtain a numerical relationship, areal values of rainfall for Denbighshire were estimated for the months November—March. Two factors were defined as follow: x = excess of November to March rainfall over 16 inches (with a minimum of 1); y = excess of accumulative (degree-month) temperature deficiency below 43°F over 10 degree-months at a reference station (Wrexham), (with a minimum of 1); the product xy was then found to be highly correlated (+0.9) with the percentage deaths of ewes carrying twins. The details are given in Table XX and illustrated in Fig. 6B.

Although this gives a reasonably satisfactory solution of the exposure effect in terms of recorded weather, it does not solve the problem of preventing the losses. Action has to be taken, in the form of extra food or temporary shelter, long before the end of the pregnancy.

Intermediate checks were therefore done on the November—January weather and the November to February weather using amended thresholds

TABLE XX

Rainfall, low temperatures and pregnancy toxaemia

Year	Rainfall	Accumulated temperature below 43° F	x	y	xy	Percentage deaths
1954—55	21.4	20.3	5.4	10.3	56	37.0
1962—63	13.1	39.8	1	29.8	30	17.5
1959—60	25.8	8.2	9.8	1	9.8	9.1
1961—62	13.7	18.8	1	8.8	8.8	8.2
1964—65	15.5	16.1	1	6.1	6.1	8.0
1963—64	11.8	17.3	1	7.3	7.3	6.8
1957—58	16.1	12.3	1	2.3	2.3	3.7
1956—57	12.7	3.0	1	1	1.0	2.6
1958—59	14.4	12.1	1	2.1	2.1	2.1
1960—61	21.2	8.0	5.2	1	5.2	1.5
1955—56	15.7	14.5	1	4.5	4.5	1.4

*For explanation of symbols see the text.

TABLE XXI

Verification of deaths from pregnancy toxaemia

Year	Forecast and *xy* values		Actual incidence	
	January 31	*February 28*		
1954—55	above average (8.7)	severe (33)	severe	correct
1962—63	severe (21)	severe (32)	severe	correct
1959—60	above average (9.6)	above average (9.6)	above average	correct
1961—62	average (6.2)	average (5.0)	above average	
1964—65	average (5.3)	average (5.9)	above average	
1963—64	average (4.7)	average (5.0)	average	
1957—58	low (1.4)	low (1.0)	low	correct
1956—57	low (1.0)	low (1.0)	low	correct
1958—59	above average (12.5)	below average (4.1)	low	
1960—61	severe (23)	average (6.5)	low	pessimistic
1955—56	low (3.4)	average (6.5)	low	

*For explanation of symbols see the text.

of 10 inches and 5 degree-months, and 14 inches and 8 degree-months, respectively. The simulated forecasts of disease loss are shown in Table XXI. The only major error was in 1960—61 when the winter started badly but improved greatly during the last two months of pregnancy.

Mycotic abortion

Work by Hugh-Jones and Austwick (1967) at the Central Veterinary Laboratory, Weybridge has shown that the incidence of mycotic abortion can be linked to the presence of a mycotoxin in hay made during wet weather. It is also interesting to note that the disease incidence is particularly severe if the animals are housed and if the hay bales are broken out in a closed building. Hay, even if it is mouldy, fed to animals in the open, with the bales opened up from the back of a farm vehicle, produces far less cases of fungal poisoning. This illustrates the importance of spore load density.

The human disease known as "farmer's lung" is caused in a similar manner. Fungi in the mouldy hay can be inhaled by farm staff and disease and death can ensue. The rainfall during hay harvest is the critical weather factor. Good well-made hay presents few problems for man or beast.

Subsequent work by P.K.C. Austwick in New Zealand showed that mycotic abortion in that country was also weather-sensitive. There the damaging fungus multiplied rapidly in the top layers of silage clamps, immediately under the covering of plastic where daytime temperatures in sunshine would be very high. The timing of the incidence of mycotic abortion in New Zealand depended on the extent to which supplementary feed in the form of silage was given to the cattle. Again the concept of "dry-days" was used, and in years when a large number of such non-growing days for grass was large, then there was an early onset of abortions. In wetter years, when less silage was fed, the incidence was up to two months later. (P.K.C. Austwick, personal communication, 1973.)

ANIMAL PRODUCTION

In investigations into problems of animal production, considerable attention has been paid to breed, although for some reason difficult to understand, conformation has been thought to be more significant than performance. Some work has been done on the effects of nutrition, but very little research has been done in regard to environmental conditions, with the possible exception of pigs and poultry.

Milk

As in many other problems, there is a great shortage of relevant data. The Milk Marketing Board of England and Wales does, however, run a milk recording scheme which gives considerable insight into the weekly and seasonal changes of milk production. When such data was examined in relation to the weather it was found:

(a) The date of spring, when the grass starts to grow and cows are turned out to graze, has a major effect on milk production in March, April and May.

(b) The April milk production is highly correlated with production over the following 10—11 months.

(c) The peak of summer milk production occurs just before the soil moisture deficit reaches 50 mm.

(d) If the deficit increases to 100 mm, a rapid fall in production is possible unless sufficient supplementary food is available.

(e) The time of the hay-harvest and especially the weather during haymaking greatly influences winter milk production.

(f) Sudden drops in temperature in late autumn will depress yields, especially if the herds are still out to grass both night and day.

(g) A drought or cold stress, unless alleviated by good dairy husbandry can have a lasting irreversible adverse effect.

Turning to the quality of milk, examination of the butter fat content, month by month over 10 years in England and Wales, shows a marked

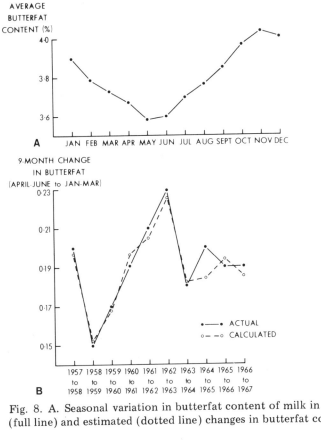

Fig. 8. A. Seasonal variation in butterfat content of milk in England and Wales. B. Actual (full line) and estimated (dotted line) changes in butterfat content of milk.

seasonal fluctuation with a maximum in November and a minimum in May (see Fig. 8A), thus being almost the mirror image of milk production per cow. Seasonal values are given in Table XXII. The largest range in quarterly change, and hence the time which is probably most sensitive to current weather is that from the 4th to the 1st quarter. This quarterly change dominates the annual change and appears to vary with June rainfall which is a good indicator of hay quality. Furthermore, dry summers appeared, on the two years when they occurred, to have an adverse effect. Measuring the England and Wales June rainfall in inches, and taking the deficiency below 2 inches as negative, the excess above 3 inches as positive, with values between 2 and 3 inches as zero to give a factor J, and taking the deficiency below 8 inches of the sum of the June, July and August rainfalls as the factor S, then the change in butter fat content can be expressed in the regression equation:

2nd quarter (Apr., May, June) to 1st quarter the next year

(Jan., Feb., Mar.) = $0.194 - 0.03J - 0.008S$

to a high degree of accuracy (multiple correlation 0.98) enabling a forecast

TABLE XXII

Seasonal variations in butterfat content of milk

Season	Mean (%)	Change from previous quarter	
		Mean	Range
1st quarter (Jan., Feb., Mar.)	3.81	−0.20	−0.25 to −0.10
2nd quarter (Apr., May, June)	3.62	−0.19	−0.23 to −0.16
3rd quarter (July, Aug., Sept.)	3.77	+ 0.15	+ 0.12 to + 0.17
4th quarter (Oct., Nov., Dec.)	4.01	+ 0.24	+ 0.21 to + 0.26

to be made at the end of August, some 7 months in advance (see Fig. 8B). The effect of June—August rainfall and hence the availability of summer grazing, was then examined in greater detail and comparing the England and Wales rainfall figures with the potential transpiration for one of the major dairying areas, Cheshire, it was found:

(a) When the potential transpiration exceeded rainfall, the change in butter fat from the 4th quarter of the previous year to the 3rd quarter of the current year (July—September), was highly correlated with the potential deficit (see Fig. 9).

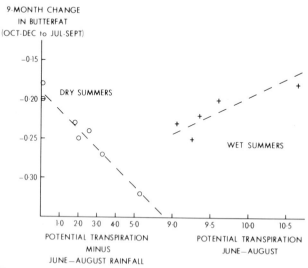

Fig. 9. Summer rainfall and the butterfat content of milk.

(b) No influence could be found on March soil temperatures, which give a good indication of date of spring and the change from January—March to April—June, but there was a striking, but unexplainable, correlation with the change April—June to July—September.

(c) When rainfall exceeded potential transpiration, the same change is negatively correlated with potential transpiration (see Fig. 9). In other words, in a dry summer, the drop is large; in wet summers when grass is plentiful the drop is smaller, especially when it is sunny.

(d) The solids-not-fat content for the "milk year", April—March, was highly correlated (0.87) with the April value (A) and the most important weather factor was again the June rainfall (J), measured as for butter fat, the formula being:

$$5.424 + 0.38A - 0.0075R$$

(e) The location of milk production is strongly related with effective transpiration, which is a powerful parameter linked to grass growth, but this is an agro-climatological exercise which has provided convincing evidence of the reliability of potential transpiration estimates.

Beef

Milk recording in Britain has been carried out for over 20 years but the corresponding beef recording scheme has only recently been started. This handicaps any attempt at a similar analysis, but C.V. Smith (1970) has done some valuable preliminary work on environmental influences. He points out that the metabolic heat generated and dissipated by an animal is a function of its size, age and weight, its food intake both in quality and quantity, its thermal environment and its degree of insulation from heat loss to such an environment.

For most animals there exists what has been termed a "comfort zone", which is probably a misnomer, and which could better be called a "thermo-neutral zone" within which the animal can adjust its body temperature without a great expenditure of energy or effort. At the lower end of this zone, there exists a critical temperature, below which the animal has to increase its heat production to maintain body temperature and use body fat for this purpose. This critical temperature, usually defined for still air conditions, will be affected by the feeding level. An animal will maintain body weight at a lower temperature level if it has high good-quality food intake.

Appraising an experiment in Ireland by McCarrick and Drennan (1972), Smith (1970) showed that during a period in 1966—67 of maintenance feeding and zero weight gain, the mean air temperature was 5.3°C. Taking this as the critical temperature, he showed that during periods of weight gain in the winter 1967—68, the mean air temperatures were 6.3°, 7.0° and 6.1°C; during periods of weight loss they were 4.8°, 2.5°, 3.2° and 2.3°C.

Smith also points out the importance of taking wind velocity and hence the advective heat loss from an animal into account. His *W.M.O. Technical Note* No. 107 gives further details concerning the use of meteorological observations in the planning and analysis of animal experiments. It is certain that a great deal more careful work has yet to be done on the quantification of the exposure factor, and no field trials should be carried out without proper environmental assessment.

Unless the weather becomes too warm so that animals seek shade and in so doing restrict their grazing, summer gains in live weight are largely a function of fodder availability. Grass growth can be linked to weather conditions with a fair degree of success so that, with homogeneous animal data, it should be possible to establish links between weather, climate and beef production. The same may be said of production of lamb and mutton.

Pigs and poultry

In Britain, pigs and poultry are for the most part, housed animals. Although a search through the agricultural literature will produce a number of papers on their living conditions and the effects of various regimes of light (for egg production), temperature and humidity, it does not appear that agricultural meteorologists in any country have done much work in this subject. This is a pity, because the assessment of physical factors, especially in regard to such aspects as ventilation, is not easily done by someone without knowledge of meteorological physics. In fact, the whole subject of internal climates is one which has for too long been neglected. The cynical comment might be made, that even so, living conditions for animals are probably better attended to than those for human beings. Life in modern glass boxes posing as offices is such that one can only assume that architects are scientifically illiterate.

Wool quality

The quality of wool in Britain, judged by the proportion which is "cast", that is to say, it is placed in the lowest category, shows remarkable variations from region to region and from one year to another. It has been suggested that this may be largely or partly due to the incidence of a disease called mycotic dermatitis, which is known to be spread during heavy rain, probably when the flocks are bunched together in search of shelter or during fright of thunder and lightning.

The Wool Marketing Board have details of such wool quality over a period of years so attempts were made to see if these were related to weather conditions. Firstly the number of thunderstorms reported at the many climatological stations were examined, to establish a mean county figure. These county figures were then weighted by the number of sheep per county to

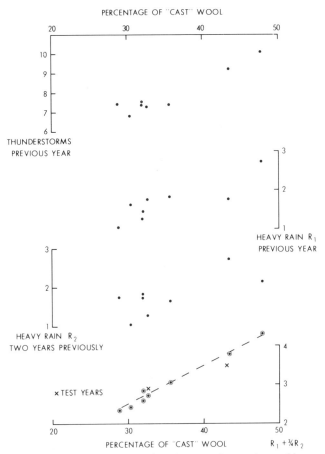

Fig. 10. Wool quality, thunderstorms and occasions of heavy rainfall.

give a regional mean comparable with the Wool Marketing Board areas. Finally a national figure was obtained by weighting these regions with the appropriate weight of wool produced.

A smaller number of more accurate observations of heavy rainfall was available for stations with recording rain-gauges. The number of occasions of heavy rain, 5 mm in 15 min, was meaned in the same manner as for thunderstorms. The results for England, Scotland and Wales are given in Table XXIII (see also Fig. 10). The correlations between cast wool and thunderstorms, and the partial correlations with the two heavy rain factors (R_1 and R_2) are very high indeed. The influence of heavy rain two years previous to shearing is interesting, and probably indicates the setting up of an infection level in the flocks. Fig. 10 shows the relation between the wool quality and $R_1 + 3/4R_2$, 1967 and 1968 being used as test years. Correlations of this kind,

TABLE XXIII

Wool quality, thunderstorms and heavy rains

Year of shearing	Percentage "cast" wool	Thunderstorm index previous year	Heavy rain index	
			previous year	two years previously
1963	28.8	7.4	1.03	1.75
1964	30.3	6.8	1.64	1.03
1965	35.6	7.4	1.82	1.64
1966	32.0	7.4	1.46	1.82
1969	47.6	10.1	2.71	2.13
1970	43.4	9.2	1.75	2.71
1971	32.0	7.5	1.28	1.75
1972	32.6	7.3	1.77	1.28

using data for a limited number of years cannot be regarded as conclusive, but they certainly encourage further investigations on the lines indicated.

International complications

It is sometimes necessary to look beyond the national boundaries to explain changes in animal production, quite apart from the international travel of air-borne pathogens. For example, some years ago, although the yield of milk per cow in England was correctly estimated, the forecast of total milk production was inaccurate because of a marked drop in cow numbers. This was due to a rise in the market price of fat cows for slaughter; this in turn was due to a decrease of imported beef from South America. The shortage of such imports was due to a drought in the Argentine. Luckily such complicated chains of cause and effect are not often encountered.

ANIMAL PESTS AND DISEASES

The discovery of the significant weather factors in the incidence and intensity of disease in farm animals is important because it leads to the introduction of a system of scientifically based warnings which enables the farmer to take evasive or curative action with consequent increase in food production. The immediate practical aim is therefore one of an approximate solution of economic importance.

Some animal diseases spread through the air, so that as in plant pathology and entomology, aerobiological considerations are the key to the epidemiology. Evidence of wind-borne spread of fowl pest was produced by C.V. Smith (1964) who examined the links between wind-directions and disease

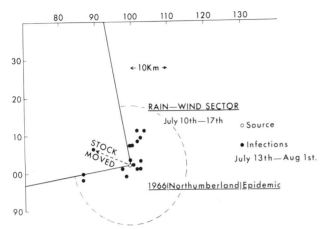

Fig. 11. Sector into which rain-bearing winds blew from an initial foot-and-mouth disease source between the time the disease started and the time it was identified, and the secondary infections which followed. (Reproduced by kind permission of the Secretary-General of the W.M.O.)

outbreaks in East Anglia. Subsequent large-scale analyses were carried out on the well-documented epidemics of foot and mouth disease which indicated the importance of aerial spread. In this disease, the data is of excellent quality and it is possible to find out the day of disease identification, the precise site and the number and type of animals involved. The length of time the disease had existed prior to identification could also be estimated with a fair degree of accuracy. It was therefore possible to examine the weather during this critical period and to establish the down-wind cone of potential air-borne infection and compare this with subsequent outbreaks. Some of the findings can be summarised as follows:

(a) Even before the statutory restrictions on animal and vehicular movement come into force, over 90% of the disease spread takes place down-wind during the period when rain is falling. (The presence of rain may imply a deposition effect or indicate humidity conditions, or both.) See Fig. 11.

(b) Almost without exception, every subsequent outbreak was within such a danger zone within a possible incubation time.

(c) The most significant time of possible aerial spread was the last 24 h before the disease was identified and the animals slaughtered; possible night-time spread seemed more frequent than in daylight hours.

(d) The decreases in epidemic intensity coincided with spells of dry weather; if calm, foggy conditions occurred, the spread was very localised.

(e) If windy, rainy periods coincided more than once with a time of maximum virus output potential, a rapid increase in disease intensity occurred.

(f) Diseased pigs appear to be a greater disease infection source than cattle; very little spread could be traced from sheep. (This was later confirmed by laboratory tests.)

(g) When large outbreaks were in existence on the neighbouring mainland of Europe and the winds were blowing towards the British Isles, scattered outbreaks occurred in the English coastal areas, which could not be traced to imported diseased meat.

These and other deductions from the weather analysis built up a mass of circumstantial evidence which suggested that the spread of foot and mouth disease by the wind was far more important than had hitherto been realized and current arrangements for the control of any future outbreak of the disease include the use of meteorological advice and mobile meteorological observing units. Later research in Switzerland and Germany has produced further evidence confirming the importance of wind-borne infection.

Stress diseases

Stress on animals may be a direct environmental effect, or it may be indirect, causing a nutritional stress, or it may be a combination of both. It is difficult to acquire adequate data on the effects of stress. Deaths, attributable at least in part to stress conditions, may be available, but the numbers of animals treated and saved from death are rarely known. Furthermore the sub-clinical effects which concern a much larger number of livestock and which may in total have an appreciable effect on production are almost unidentifiable.

The best documented losses are those met with in transit (see p. 155), and during pregnancy or at birth (see p. 103), but nutritional deficiency diseases such as swayback (a copper deficiency) and hypomagnesia can be shown to be influenced by the weather (see Fig. 12).

The reduction of environmental stress on animals is a prime requirement of increased efficiency and involves problems of shelter, housing and animal behaviour. With an increasing demand for food and better energy conversion rates, there is still a great deal to be learnt in this problem.

There may be occasions when a succession of stresses have to be identified. Recent research indicates that the incidence of acute respiratory disease or pneumonia in young animals may be influenced by the occurrence of a cold stress at the time of infection, followed by a second cold stress coinciding with the end of the incubation period of the pathogen. This corresponds to the conditions leading to the rapid build-up of foot and mouth disease, namely a double coincidence of weather and disease factors, and may indeed be of major fundamental importance in the epidemiology of many diseases.

Critical weather factors

The identification of a critical weather factor in a disease can be of the

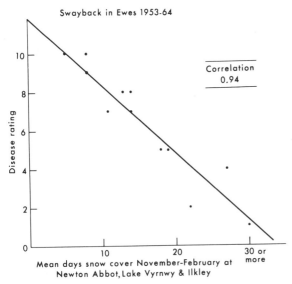

Fig. 12. National severity (scale 0—10) of swayback in ewes and days of snow-cover during the winter during which supplementary feeding was likely, thus reducing the risk of a copper deficiency. (Reproduced by kind permission of the Secretary-General of the W.M.O.)

greatest importance. As in plant disease, fungal diseases of animals lend themselves to this form of appreciation and interpretation. Although ideally knowledge is required regarding the complete progress of the disease, the identification of a factor producing a "trigger-effect", specified in terms of weather conditions, can go a long way towards providing a practical answer in the search for a system of disease forecasting. Examples of this are found in facial eczema of sheep, which was first of importance in New Zealand, but has recently been reported as occurring in South Africa, and which is due to a fungus on grass. Another example is mycotic abortion, which is due to mouldy hay or silage; mycotic dermatitis, which seriously affects the wool quality has been discussed on p. 110.

The critical weather factor need not have occurred in the immediate past, it can at times be identified several months before the onset of the disease, and this is especially true in the case of many parasitic diseases.

Parasitic diseases of animals

There are four main avenues of approach in the problem of determining the physical conditions which affect animal parasites, none of which are sufficient in themselves, but each of which has something to offer towards the solution of the problem. These are laboratory experiments, plot trials, field observations and regional disease statistics.

Work in a laboratory is attractive, if only for the fact that the building and facilities are there, staff and money, although limited, are available, and work materials can be brought in. Moreover, much outstanding scientific work has been produced under such conditions. There are, however, certain hidden pitfalls. It is very difficult indeed to reproduce outside conditions and all their fluctuations in time and space within controlled artificial environments. The very impression of control may be illusory. As a result the interpretation of laboratory results in terms of outdoor reality and scale is fraught with difficulty and sometimes leads to baffling contradicitions.

A step towards natural conditions involves the use of small experimental plots, in which the circumstances correspond more closely to those in field pastures. Difficulties, however, still remain, for it is difficult to reproduce the behaviour of the grazing animal and an important factor in the disease build-up may be wrongly deduced. The next step is to work with experimental small fields or paddocks, in which the effect of hard management or fodder control can be simulated.

A survey system of field observations abandons a degree of control but gains a representation of reality. The difficulties arise because of the need for adequate sampling in time and locality and the consequent high demand on manpower and mobility. Finally the use of data of disease incidence, and thus to a certain extent, working backwards, again demands a reliable homogeneous reporting system which may not exist except in limited localities. In the end, however, such "end-point" details are essential before practical, as distinct from theoretical, progress is made.

The point to remember is that each approach is guided by results obtained by other methods, and conversely they provide results which can be fed back into the collateral research. The urgent demand of science is to explain observed phenomena; the urgent demand from agriculture is for advice to help combat disease losses. Neither demand can be satisfied by work in isolation; each type of research worker is contributing to the composite problem.

Fascioliasis

Attempts to discover the weather factors relating to any disease are made easier if there is available a long history of the disease with a fairly accurate knowledge of the times of serious outbreaks. A case in point is fascioliasis, or liver fluke, which has been recognized as a serious danger to sheep and cattle for at least 200 years. Traditional knowledge, never to be ignored or underestimated in any scientific investigation because the facts are likely to be near the truth, even if the reasoning is faulty, has always associated heavy stock losses with wet summers.

This can easily be verified by a reference to the monthly percentages of average rainfall for the country as a whole during the years when records show that serious outbreaks of the disease occurred. In England and Wales

TABLE XXIV

Mean monthly rainfall during years with serious liver fluke disease (expressed as percentages of longterm average)

January	105	May	115	September	142
February	102	June	143	October	86
March	80	July	123	November	101
April	113	August	105	December	102

there were 36 such years in the period 1734—1960; during these years the mean percentages of rainfall average were as shown in Table XXIV. This confirms the general opinion that rainfall in summer is an important factor, with the suggestion that June and September are critical months.

Records furthermore show that outstanding losses occur in seven years, namely 1830, 1860, 1879, 1924, 1946, 1954 and 1958. The mean percentages during these years are shown in Table XXV.

It is now clear that rainfall in summer and early autumn could be regarded as a dominant weather factor, but a far more detailed analysis is necessary before any major progress can be made. Percentage of average of an areal value is a crude enough parameter in any case, but if wetness of pasture is significant, the rainfall must be considered with reference to the drying process of evaporation.

At this stage, the direction of research depends greatly on the extent of knowledge of the life-cycle of the parasite. In the case of liver fluke, a great deal is known about *Fasciola hepatica* and its intermediate host, the snail *Hymnaea trunculata*. For example, laboratory experiments and field experience have shown that there is no development of the egg of a parasite below 10°C and that the process of hatching may take several weeks within the 10°—20°C temperature range. This clearly puts a northern climatic limit on the areas likely to be affected by the disease and it does not occur in northern Norway or Iceland even though infected sheep are occasionally introduced and the necessary snail host is known to exist in these regions.

Experience has shown that over much of the British Isles, the hatching of

TABLE XXV

Mean monthly rainfall during years with severe liver fluke diseases (expressed as percentages of longterm average)

January	106	May	148	September	159
February	117	June	175	October	77
March	75	July	122	November	105
April	95	August	141	December	92

eggs deposited on pasture via the dung of infected animals takes place during the period May—October, the bulk occurring between June and September during whcih time the soil temperatures almost always exceed the 10°C limit. In northern Scotland conditions are cooler and the completion of the life-cycle in one season is difficult. Dry conditions also retard the hatching of the eggs but this is not a major factor with respect to disease spread as there is always a continual flow of viable eggs being deposited.

After hatching, the parasite must quickly find the intermediate snail host and it is likely that it is the relative prevalence of the snail that determines the intensity of subsequent disease, for if no snail is found, the life-cycle ends. These snails live in grass habitats which are moist but not too acid; in winter there are no problems as with few occasions rainfall exceeds evaporation; in summer the duration of wet grass conditions is critical. It is probable that this is the simplest expalantion of the overwhelming importance of summer rainfall. Snail populations increase rapidly in favourable conditions and the more they thrive then the larger the parasite population they can support.

It was therefore desirable to obtain from weather records an expression which would give a measure of the time the grass was wet. The selected parameter was $(R - P + 5)n$, where R = monthly rainfall in inches; P = monthly potential transpiration in inches; n = number of rain days per month. Field observations showed that when this parameter reached 100, then the habitats were wet throughout the month, and as it is impossible to be wetter than wet in this context, a maximum of 100 was imposed. To allow for the influence of temperature the monthly values for May and October were halved. The maximum rating for a very wet summer was therefore:

50 (May) + 100 (June) + 100 (July) + 100 (Aug.) + 100 (Sept.) + 50 (Oct.) = 500

Detailed records of disease incidence, which occurs from late autumn onwards, after the parasite *Cercariae* had left the snail and infected the herbage were available for Anglesey for the years 1948—1957. It was therefore possible to compare these with the meteorological parameter, and it was found that losses from acute disease did not occur until its value reached 300; the losses were above average above 400, whilst severe outbreaks occurred at 475.

Forecasts of disease intensity, based on this model were first issued in 1958, an outstandingly bad year for losses; the following dry year was the complete reverse and these two initial successes led to a regular issue of liver fluke warnings on the broad scale of low, below average, above average and severe for the several regions of the British Isles, making use of the monthly meteorological records from over 100 reporting stations.

The reliance on standard meteorological observations can be weakened in areas with a high water-table (as in parts of The Netherlands) or in regions where large-scale irrigation is practiced. Nevertheless, experience has shown that analyses on these lines have been successful in identifying years of high disease risk in several European countries.

The important practical aspect of this system is that advice can be given to farmers in time for them to take action before the herbage becomes infected and disease losses commence. To issue a forecast at the end of August does imply some assumptions regarding September and October weather, but otherwise the decisions are taken by reference to events that have already taken place.

Nematodiriasis

This disease of young lambs is due to a single species of nematoid, *Nematodirus battus*, and is restricted to a limited season in late spring. Its life-cycle is known in general terms, in that infected lambs deposit the parasite infection in their faeces in early summer, the eggs subsequently developing slowly through the successive larval stages reaching an unhatched infective stage in autumn. The critical factor is the time of hatch in the rising temperatures of the following spring. If this hatch is relatively late it will coincide with the time when the next year's lambs begin to graze and they will become infected and possibly die. If the hatch is early, infection and losses are low.

The obvious meteorological parameter to consider is soil temperature and this was confirmed when it was found that the mean March earth temperature under grass at 30 cm depth from even a single climatological station (see Fig. 12) gave a good indication of the nation-wide severity of the disease in April and May.

Using such temperatures for two stations, Oxford in central England (T_1) and Cockle Park in the northeast (T_2) an index was developed:

$$52.5 - T_1 - T_2/10 \text{ where } T_1, T_2 \text{ are in } °F$$

If this index was $\geqslant 8$ a high disease incidence was expected; 5–8 above average; 3–5 below average; and < 3 a low incidence.

Over a period of years this semi-empirical forecast aid gave very satisfactory and useful forecasts with one exception, in 1954, when disease losses were much higher than expected. April that year was unusually dry (see Fig. 12), which indicated that although in spring in most years the soil-moisture conditions were sufficiently moist to have only a minor effect on hatching and that soil temperature was the dominant factor, there were occasions when abnormal weather would decrease the accuracy of any simple approach.

This is an illustration of a fact of life that a worker with field data is restricted to one experiment or one test per year. Unless his years of trial

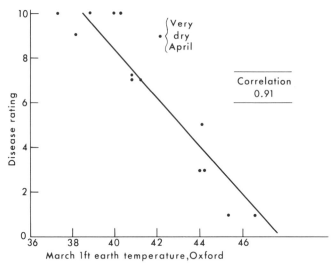

Fig. 13. National severity (scale 0—10) of nematodiriasis in lambs and March earth temperatures. (Reproduced by kind permission of the Secretary-General of the W.M.O.)

contain a fair sample of possible weathers, he is always liable to overlook the significant factor of an unusual year, simply because he so far has not analysed the effect of such conditions.

Clearly, a big step forward has been made by the intelligent use of the very minimum of meteorological information, but it is only a first step. In most years, and especially in northern England, the hatch of the nematoids is confined to a limited period of days and rises to a very definite peak. Knowledge of this peak hatch date would be vary helpful in knowing when to keep new lambs off pastures which had been contaminated the previous year.

Work by Thomas (1972) at Newcastle University provided data on the times of hatching from several different sites for an 11-year period. It was then found possible to derive a formula for the peak hatch date using the mean 30 cm earth temperature (T_{20} °F) over the March 1st—March 20th period, namely:

Number of days after March 31st to peak hatch data = $182 - 4T_{20}$.

The errors introduced by the use of such a formula rarely exceeded 7 days, and such errors could be reduced by taking into account any dry days which occurred up to the hatch date. As advice was needed by the third week in March, such corrections could not be incorporated in any forecast but they serve to show that both temperature and moisture are important in considering the process of hatching.

Ostertagia and Trichostrongylus infections

The importance of soil moisture had already been recognized in regard to the incidence of parasitic gastroenteritis in sheep and cattle. It was found that high soil deficits led to a marked decrease in disease in late summer, but that dry autumns seemed to be followed by a marked increase in the moist conditions of the following May.

If sheep and lambs are turned out onto clean pasture, the only source of infection for the lambs is the post-parturient rise in egg deposition by the ewes. This deposition occurs at approximately the same time each year so that variations in peak hatch dates obtained from pasture larval counts must be largely influenced by the intervening weather; in wet years the peak hatch is early in July, in dry years it is much later, at times even occurring in September, irrespective of the variations in the temperature regime.

Using records of 12-h totals of rainfall it is possible to estimate how long the grass has been wet in terms of 6-h periods, due account being taken of evaporation rates and time of day of the rain. Recent investigations have shown that egg and larval development only takes place in relative humidities near 100%, and the limited series of field experiments suggests that some 600 h of wet grass are needed.

Babesiosis (Redwater fever)

The two previous examples of investigations into the influence of weather on parasitic diseases of animals have for the most part concentrated on the effects of the current season, leaving out any consideration of previous seasons which could have affected the parasite population. This is not always advisable because critical weather-sensitive events may have occurred in the early stages of the life-cycle of the parasite or its intermediate host.

Redwater fever is caused by a parasite *Babesia divergens* which is vectored by the common tick *Ixodes vicinus*. In areas with suitable habitats for the tick the values of disease incidence were as high as 1/2 or 3/4 of the cattle (sheep are immune) at risk per month, mainly during the period April—October; the disease can be fatal.

Donelly and MacKellar (1970) have investigated this disease in detail in South Devon. They compared measurements of tick population in the field with disease incidence and found there was a high correlation between the two provided that they incorporated a 2½ week time lag in the disease figures. This delay is explained by the time needed for activation and host-seeking by the tick and the development of the effect of the parasite on the animal producing clinical symptoms.

They furthermore showed that the disease prevalence was highly correlated with maximum air temperatures occurring 14 days earlier. The discovery of this weather parameter is interesting because it implies two things; firstly that the weather is affecting the vector tick and secondly that maximum air temperature is a better simple indicator of the significant temperature in the tick habitat (in the grass) than mean air or minimum air temperature.

This temperature correlation fell into two phases; the January—May and the June—December data had to be considered separately. They found the following (statistically significant) regression equations:

January—May $D = 14.45T - 105.6$
June—December $D = 6.2T - 48.1$

where T = mean monthly maximum air temperature; and D = disease per 30—31 days starting 14 days after beginning of the month to which T refers.

The cause for the large difference in temperature coefficient has to be sought in the physiological state of the tick and the time it completed its engorgement (blood-feeding) the previous season.

High correlations were also found with mean monthly soil temperature, day length and sunshine hours, all of which are themselves strongly related to maximum air temperature. A small negative correlation was found with rainfall totals. A dry month may cause a drop in grass growth, forcing the grazing animal to eat less selectively and become at greater risk to tick infection. It was clear, however, that this was of second-order importance and the dominant weather factor on the (cold-blooded) tick was the maximum air temperature.

The disease incidence is thus bimodal, and can be divided into spring and late summer-autumn totals for the purpose of investigating the effect of tick population in preceding seasons. Recent research gave the following results.

Spring totals

The simplest weather parameter of mean May maximum air temperature (at Newton Abbot) was chosen and it was found for 12 years, 1955—1967:
Total correlations:

spring total	and May mean maximum	+0.4584
spring total	and previous autumn total	+0.4342
previous autumn	and May maxima	−0.3518

giving partial correlations:

spring total	and May mean maximum	+0.73
spring total	and previous autumn total	+0.72

with a double regression:
spring disease total = $32T_m + 0.5A - 434$ (see Table XXVI) where T_m = May mean maximum temperature in °C; A = disease total the previous autumn.

This analysis was then extended further to see whether the previous spring totals had any effect. The partial correlations then became:

spring total	and May mean maximum	+0.7420
spring total	and previous autumn	+0.7333
spring total	and previous spring	+0.0770

TABLE XXVI

Verification of estimates of redwater disease in spring

Year	Previous autumn total	Mean May maxima (°C)	Estimate	Actual disease	Error
1956	332	17.7	298	316	−18
1957	321	15.6	226	240	−14
1958	438	15.0	265	216	+49
1959	260	17.1	243	189	+54
1960	329	17.0	275	295	−20
1961	324	16.1	243	216	+27
1962	461	13.8	238	230	+ 8
1963	341	15.0	216	234	−18
1964	323	16.3	248	274	−26
1965	329	15.1	214	270	−56
1966	209	15.4	164	152	+12
1967	240	14.8	160	152	+ 8
				Mean	26

Thus the previous spring appears to have no effect at all, and this discredits the widely-held theory that there are two distinct behavioral patterns of tick, one concerned with spring activity and the other with autumn.

Autumn totals

In this case the simplest weather parameter was the sum of the June and July mean maximum air temperatures. It was found that the partial correlations, considering only the previous spring totals, were:

autumn total and June—July maxima   +0.68
autumn total and previous spring total   +0.56

The inclusion of the previous autumn total, as well as the previous spring produced a result different to that found in the case of the spring. The partial correlations then became:

autumn total and June—July maxima (T_{JJ})   +0.82
autumn total and previous spring totals (S)   +0.42
autumn total and previous autumn totals (A)   +0.66

This autumn—autumn effect is confirmed by experimental work in Czechoslovakia which showed that ticks which do not complete their engorgement until late in the year, take 9 months or more to hatch. The triple regression became:

autumn disease total = $44.2 T_{JJ} + 0.69A + 0.44S − 1753$ (see Table XXVII).

TABLE XXVII

Verification of estimates of redwater disease in autumn

Year	Previous autumn total	Previous spring total	June + July mean maxima ($°C$)	Estimate	Actual	Error
1956	332	316	38.3	309	321	−12
1957	321	240	42.1	436	438	− 2
1958	438	216	37.5	303	260	+43
1959	260	189	41.6	349	329	+20
1960	329	295	40.3	386	324	+62
1961	324	216	40.8	370	461	−91
1962	461	230	37.8	338	341	− 3
1963	341	234	37.9	262	323	−61
1964	323	274	39.2	324	329	− 5
1965	329	270	36.7	216	209	+ 7
1966	209	152	39.8	218	240	−22
1967	240	152	40.0	249	191	+58

Mean   32

REFERENCES

Hugh-Jones, M.E. and Austwick, P.K.C., 1967. Epidemiological studies on bovine mycotic abortion. *Vet. Rec.*, 80: 273—276.
McCarrick, R.B. and Drennan, M.J., 1972. Effect of winter environment on growth of young beef cattle. *Anim. Prod.*, 14: 97—105.

SUGGESTIONS FOR FURTHER READING

Animal reproduction

Ollerenshaw, C.B., Smith, L.P. and Michael, D.T., 1972. A relationship between climate and the incidence of pregnancy toxaemia in sheep in the Denbighshire area of Wales. *Biometeorology*, 5 (Part 1): 92.

Animal production

Gloyne, R.W., 1964. Climate and the blackface sheep industry. *J. Blackface Sheep Breeders Assoc.*, 17: 29—33.
Gloyne, R.W., 1964. Weather of recent seasons in the blackface sheep areas. *J. Blackface Sheep Breeders Assoc.*, 17: 1—4.
Smith, L.P., 1965. Changes in the distribution of milk production in England and Wales. *Outlook on Agric.*, 4: 302—309.
Gloyne, R.W., 1967. A contribution on methods of investigating the effects of weather and climate on farm livestock under normal management in the British Isles. *Int. J. Biometeorol.*, 2 (part 1): 387—394.

Smith, L.P., 1968. Forecasting annual milk yields. *Agric. Meteorol.*, 5: 209—214.

Smith, C.V., 1970. Meteorological observations in animal experiments. *W.M.O. Tech. Note*, No. 107, 37 pp. (see Appendix I).

Animal pests and diseases

Ollerenshaw, C.B., 1959. The ecology of the liver fluke, *Fasciola hepatica. Vet. Rec.*, 71: 957—963.

Ollerenshaw, C.B. and Rowlands, W.T., 1959. A method of forecasting the incidence of fascioliasis in Anglesey. *Vet. Rec.*, 71: 591—598.

Smith, L.P., 1963. A fluke, a snail and the weather. *W.M.O. Bull.*, 12: 198—200.

Smith, C.V., 1964. Some evidence of the windborne spread of fowl pest. *Meteorol. Mag., Lond.*, 93: 257—263.

Ollerenshaw, C.B., 1966. The approach to forecasting the incidence of fascioliasis over England and Wales. *Agric. Meteorol.*, 3 (1/2): 35—54.

Ollerenshaw, C.B. and Smith, L.P., 1966. An empirical approach to forecasting the incidence of nematodiriasis over England and Wales. *Vet. Rec.*, 79: 536—540.

Smith, L.P. and Ollerenshaw, C.B., 1967. Climate and animal disease. Recent developments in forecasting the incidence of four livestock diseases in Britain. *Agric., Lond.*, 74: 256—260.

Hurst, G.W., 1968. Foot and mouth disease. The possibility of continental sources of the virus in England in epidemics of October 1967 and several other years. *Vet. Rec.*, 81: 610—614.

Ollerenshaw, C.B., 1969. A relationship between the incidence of swayback in England and Wales and its application to forecasting the incidence of the disease. *Biometeorology*, 4 (Part II): 100.

Ollerenshaw, C.B. and Smith, L.P., 1969. Meteorological factors and forecasts of helminthic disease. *Adv. Parasitol.*, 7: 283—323.

Smith, L.P. and Hugh-Jones, M.E., 1969. The effect of wind and rain on the spread of foot and mouth disease. *Nature, Lond.*, 223: 712—715.

Smith, L.P. and Ollerenshaw, C.B., 1969. The role of weather in animal disease problems. *Vet. Ann.*, 10: 286—296.

Wright, P.B., 1969. Effects of wind and precipitation on foot and mouth disease. *Weather, Lond.*, 24: 204—213.

Donelly, J. and MacKellar, J.C., 1970. The effect of weather and season on the incidence of Redwater fever in cattle in Britain. *Agric. Meteorol.*, 7 (1): 5—18.

Hugh-Jones, M.E. and Wright, P.B., 1970. Studies on the 1967—8 foot and mouth disease epidemic. *J. Hygiene*, 68: 253—271.

Smith, L.P., 1970. Weather and animal disease. *W.M.O. Tech. Note*, 113 (see Appendix I).

Smith, L.P. and Thomas, R.J., 1972. Forecasting the spring hatch of *Nematodirus battus* by the use of soil temperature data. *Vet. Rec.*, 90: 388—392.

Thomas, R.J., 1972. Forecasting the spring hatch of *Nematodirus battus*, a nematode parasite of sheep. *Biometeorology*, 5 (Part 1): 113.

Ollerenshaw, C.B., 1974. Forecasting liver-fluke disease. *Symp. Brit. Soc. Parasitol.*, 12: 33—52.

Smith, L.P., 1974. Meteorological factors of importance in biological systems. *Symp. Brit. Soc. Parasitol.*, 12: 1—12.

Thomas, R.J., 1974. The role of climate in the epidemiology of nematode parasitism in ruminants. *Symp. Brit. Soc. Parasitol.*, 12: 13—32.

CHAPTER 5

The Modes of Agricultural Meteorology — Soils

> "Error is a hardy plant: it flourisheth in
> every soil."
> M. TUPPER

SOIL CONDITIONS AND DRAINAGE

Soils, of all types, are in a much more suitable condition for ploughing and other cultivations when their moisture content is below field capacity. If they contain excess water they are much more likely to suffer structural harm when they are traversed by heavy machinery. Once the soil has reached a state of field capacity the number of days when it is fit to be worked are few so that the date of return to capacity in the autumn is of major importance to arable husbandry. Indeed, one of the proofs of the accuracy of the calculated return dates is the close relationship they have with the area of land sown to wheat, which is chiefly an autumn-sown crop.

The average return dates for various areas of England and Wales can be found, and the percentage of land under wheat is given in official statistics. The details are given in Table XXVIII.

Average values may be misleading, so that year-to-year variations in acreage must be considered to show that the return date is a true and valid integrator of the time available to carry out the autumn work. Accurate information was available for two regions in the West Midlands of England, one with a clay, and the other with a sandy soil. Return dates were calculat-

TABLE XXVIII

Average return to capacity dates and preference for wheat

Region	Average return date	Percentage of farmland under wheat	
		1959	1967
Wales	Aug. 30	0.7	0.9
N. England	Sept. 24	2.8	2.7
S.W. England	Oct. 10	2.9	4.6
Yorkshire Lancashire	Oct. 11	8.1	7.1
W. Midlands	Oct. 21	6.3	7.7
S.E. England	Nov. 6	8.6	11.4
E. Midlands	Nov. 11	12.5	14.7
E. England	Dec. 1	16.7	19.7

TABLE XXIX

Return-to-capacity dates and wheat areas

Year	Return date	Percentage wheat	Year	Return date	Percentage wheat
Clay soil			Sandy soil		
1964—5	Mid-January*	60	1964—5	late-December	36
1962—3	late-December	51	1969—70	early-December	30
1959—60	mid-December	53	1966—7	early-November	28
1961—2	mid-December	50	1967—8	late-October	31
1969—70	late-November	46	1968—9	late-October	27
1963—4	mid-November	51	1965—6	late-September	23
1965—6	early-November	49			
1967—8	early-November	48			
1966—7	late-October	49			
1958—9	early-October	43			
1960—1	early-October	41			
1968—9	early-September	34			

*The first 10 days of a month are called "early"; the second 10 days "mid"; the last 10—11 days "late".

ed from records of local rainfall and potential transpiration estimates. The results are shown in Table XXIX and illustrated in Fig. 14. On both soils, there is clearly a high correlation between the return date and the wheat area. The proportion of land in wheat on the heavier soil is greater because the thinner sandy soil is less suitable for the crop. The sensitivity of the heavier soils to moisture conditions was to be expected, but this analysis shows than even the lighter soils, thought to be more easily workable in wet conditions, are affected by the current weather and that the cereal acreages are largely determined by the availability of days suitable for soil cultivations.

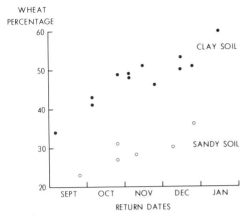

Fig. 14. Return-to-capacity dates in the autumn and the proportion of cereal land in wheat the following year.

The relationship holds true on a national average basis, it holds true also for a locality on a yearly basis; it remains to show that it is also reliable nationally on a yearly basis. The percentage of cereal land in wheat is known, but not the proportion that is spring sown. However, if an areal mean return date is found by weighting sample dates with respect to the cereal area in their vicinity, then for England, the results shown in Table XXX are obtained.

The year 1962—1963 is an anomaly because of the severity of the winter which completely altered the normal pattern of spring sowing; little spring wheat was sown and almost all the spring sowings were of barley.

It is also interesting to note that the return date in autumn appears to

TABLE XXX

National return dates and wheat areas

Year	Mean national return date	National wheat (percentage)
1964—65	January 22	31.8
1962—63	December 16	27.2
1959—60	December 9	31.4
1961—62	November 26	32.6
1963—64	November 20	29.5
1967—68	November 2	28.8
1966—67	October 25	27.4
1965—66	October 25	26.9
1960—61	October 7	27.3
1968—69	September 13	24.4

exercise a strong control on the area of land which is left fallow the following year; the spring weather seems to have far less influence. This suggests that time lost due to bad weather in autumn cannot be made up in a full spring programme. In the sequence of years quoted for wheat percentages, the millions of hectares left fallow were: 0.6, 0.8, 0.7, 0.65, 0.7, 0.7, 0.85, 1.0, 1.15 and 1.6. It must be concluded that a return to capacity before the end of October will result in an increase in fallow area the following summer.

A return date does not really provide a forecast of the sowing of winter wheat because it occurs after the event. The timing of this return to capacity depends mainly on two weather factors, namely the soil moisture deficit at the end of summer and the rainfall in autumn. Forecasts of future rainfall amounts are completely unreliable, but the deficit is known, and it can be used as the basis of a true forecast.

If the national mean soil moisture deficit at the end of September is found by weighting sample deficits with respect to the appropriate cereal acreage (as for the mean return date), then it is found that its correlation with the subsequent wheat percentage was 0.92 over the 1960—69 decade, the regression equation being $24.4 + 0.061d\%$ in wheat, where d = mean deficit in millimetres (see Table XXXI).

The soil-moisture conditions at the end of summer are therefore a reliable basis for an estimate of the amount of time available in autumn to carry out cultivations and sowing, and the wheat acreages of the following year can be foretold with some accuracy (see Fig. 15).

This, however, is a generalized conclusion; it does not take into account the response of individual soils. If an appreciable rainfall adds moisture to the soil without bringing it back to capacity, then the surface layers will be

TABLE XXXI

Autumn soil moisture deficites and wheat areas

Year	Soil moisture deficit Oct. 1st	Percent wheat following year	Estimated percentage
1959	125	31.4	32.1
1960	58	27.3	27.9
1961	102	32.6	30.7
1962	65	27.2	28.4
1963	78	29.5	29.2
1964	112	31.8	31.3
1965	18	26.9	25.5
1966	63	27.4	28.2
1967	78	28.7	29.2
1968	3	24.4	24.6

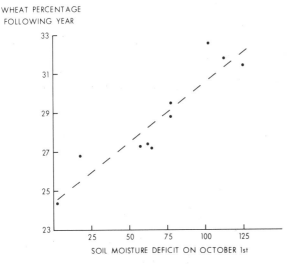

Fig. 15. Soil moisture deficits in autumn and proportion of cereal land in wheat the following year.

saturated, and a certain degree of drying is necessary before cultivations can safely start. This delay will vary with soil type, with the natural drainage, and with the moisture holding capacity of the topsoil. To specify the length of such a delay, a model can be devised as shown in Table XXXII. The variation with month is an attempt to assess the diminishing drying power of the weather; the variation with drainage category is arbitrary, but probably covers the range of behaviour of most farm soils.

Using this model, the daily autumn rainfalls, up to the date of return to field capacity were analysed for two stations over a 20-year period. Counting from the beginning of September, allowing for the delays indicated from the model, the potential work days (including Sundays) were found for each category of return date. Meaning the results within such categories the value

TABLE XXXII

Delay period in days after rain (soil below capacity)

Month	Drainage category			
	good	moderate	poor	bad
Sept.	0	½	1	1½
Oct.	½	1	1½	2
Nov.	1½	2	2½	3
Dec.	2	2½	3	3½

TABLE XXXIII

Mean work-days and return-to-capacity dates

Return date	Mean work days in various drainage categories			
	good	moderate	poor	bad
Early-Sept.	5	5	4	3
Mid-Sept.	6	5	4	3
Late-Sept.	13	12	10	8
Early-Oct.	16	14	12	10
Mid-Oct.	25	22	19	16
Late-Oct.	26	23	20	17
Early-Nov.	31	27	24	21
Mid-Nov.	36	32	28	25
Late-Nov.	35	29	25	21
Early-Dec.	45	40	36	33
Mid-Dec.	48	45	41	38
Late-Dec.	59	53	48	44

of the return date as an integrator of autumn conditions becomes obvious (see Table XXXIII).

The correlation between return date and available work days is very high in all the drainage categories, and to a first approximation, the categories described as moderate, poor and bad are 10%, 20% and 30% worse than the soils described as good (see Fig. 16).

Drainage, whether natural or artificial, becomes a much more important factor after the soil has returned to capacity. This is especially the case in spring when the soil goes through a series of dry—wet, wet—dry sequences

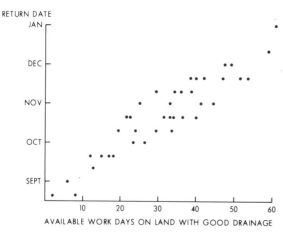

Fig. 16. Return-to-capacity dates and available work days in autumn.

following spells of rainy or rainless weather. It is in spring that the delay-days, the waiting time for the soil to become suitable for cultivations, are critically important because if some crops are planted too late, their potential yield is correspondingly reduced.

In late winter and early spring any rain is likely to fall on soil already at field capacity, so that this excess rain has to be drained away; the time taken to do this is a function of the amount of excess soil water and the drainage characteristics.

To estimate the delay occasioned by this drainage process, a second model of delay days has to be evolved. Once this has been done, it is possible to compute the total length of time needed to complete a given schedule of operations on the land in any given sequence of weather. The model for delay days when the soil is at capacity was devised as shown in Table XXXIV.

Very few soils have better drainage characteristics than those labelled "good", and very few are worse than those labelled "bad". Those soils with quicker drainage would be too thin to grow any worthwhile crop, and those with slower drainage would be so stiff and unmanageable that they would be totally unsuitable for arable farming and would probably be left in permanent pasture.

Spring cultivations and sowing must be fitted in between the beginning of March (or late February if conditions are especially favourable) and the middle of May. Using this model it is now possible to find the distribution of work days within this period over a period of years. This can be done for each category of drainage and for work schedules of varying length such as 7, 10 or 14 days.

Analyses of this sort will soon show that it is the sequence of dry-days which is important and not the total rainfall of the period. For example, imagine three sequences of weather:

Sequence 1:   60 mm rain, followed by 12 dry days;

TABLE XXXIV

Drainage categories, excess spring rain and delay days

Amount of excess water to be drained away	Drainage category (delay days)			
	good	moderate	poor	bad
Under 5 mm	1	1	2	2
5—25 mm	1	2	3	4
25—50 mm	2	3	4	5
Over 50 mm	2	4	5	6

Sequence 2:   30 mm rain, followed by 6 dry days, then
                30 mm rain, followed by another 6 dry days;
Sequence 3:   20 mm rain, followed by 4 dry days, this being repeated a
                further two occasions.

In each case the total rainfall is 60 mm, and the total number of dry days is 12, but the net effect on the available work days is very different, as is shown by the following analysis.

Calling the good, moderate, poor and bad drainage conditions, A, B, C, and D, respectively, the effect is shown in Table XXXV.

Thus good drainage enables a soil to be worked in each of the three types of weather; moderate drainage implies only a small hindrance to work in broken weather; poor drainage will only permit work to be done a short period at a time; bad drainage needs long dry spells of weather before any work can be accomplished.

To make use of such analyses of weather sequences and to understand the significance of delays, it is necessary to know the effect of late sowing on the final yield of the crop. Once this relationship is known it is possible to convert the established delays in completion of work schedules into terms of yield loss. Knowing how such losses vary with respect to length of schedule and drainage category, it is then easy to decide whether it is more economic to buy more machinery to try and reduce the schedule length, or to install better drainage to reduce the delays.

There are other times in the life of a crop when good drainage is necessary to reduce the time when plant roots are in saturated or waterlogged soil. Moreover, a well-drained soil encourages the development of a good root system, thus improving the growth and giving the plant a better chance to

TABLE XXXV

Effect of rainfall distribution on available work days under various drainages

Weather sequence	Delay days				Work days				Total dry days
	A.	B.	C.	D.	A.	B.	C.	D.	
1. (one dry spell)	2	4	5	6	10	8	7	6	12
2. (two dry spells)	2	3	4	5	4	3	2	1	6
	2	3	4	5	4	3	2	1	6
				total	8	6	4	2	12
3. (three dry spells)	1	2	3	4	3	2	1	0	4
	1	2	3	4	3	2	1	0	4
	1	2	3	4	3	2	1	0	4
				total	9	6	3	0	12

resist drought, but it is difficult to assess these benefits from meteorological data.

An agro-climatological assessment of the risks of periods of heavy rain can help to design the most suitable drainage system which has to be incorporated into a soil which is naturally poor in drainage, but it is doubtful whether any additional mole or tile drains will improve a soil to any extent greater than one shift of category; in other words, D to C, C to B, or B to A.

FERTILIZERS

Many experiments on the effect of fertilizers have been carried out in several countries on a variety of crops. Few published accounts of such experiments are found to contain adequate information of the weather conditions which prevailed. Sometimes rainfall totals are quoted, but it is rarely possible to ascertain the distribution of the raindays, and transpiration estimates are almost invariably missing; any mention of the weather during the season preceding the experiment is likely to be only in the most general terms. This is a most unsatisfactory state of affairs.

Although the reports will quote how much of the fertilizer, nitrogen for example, has been applied to the soil, they will very seldom indicate the level of nitrogen in the soil before such an application. The reader thus knows how much has been added but not the sum total, because the initial level is unknown. The extent to which nitrogen will remain in the soil over a wet winter period will depend on how much leaching has occurred, that is to say, how much excess winter rain has passed through the soil into deep seepage or drainage and carried soluble nitrogen with it.

The availability of nitrogen compounds to a plant via its roots is a function of the soil temperature, but also depends on the level in the soil reached by the additional fertilizer which is affected by the downward movement of water. An early application of nitrogen followed by a spell of wet weather may result in much of the fertilizer being washed down below the root level of young plants. A late top-dressing, followed by a dry spell, may mean that none of it reaches the root zone for a long time and therefore have little effect on growth. A complete understanding of the effect of fertilizer additions therefore needs a close analysis of the soil moisture regime which is a complex process involving detailed knowledge of rainfall and evaporation.

Some idea of the effect of adding plant nutrients can be obtained by the use of macro-scale data. On p. 78 reference was made to the dependence of hay yields in England on a weather factor (effective transpiration) and an improvement factor (assumed to be linear). There are good reasons to suppose that this improvement in grassland husbandry is largely an increased use of nitrogenous fertilizers. The total amount of nitrogen applied to grasslands in England is known on a yearly basis. If it is assumed that this bears a constant relationship to the amount used on grass grown for hay, probably true within a 10% range of error, then the hay yields can be re-analysed to obtain the following equation:

Yield (kg/ha) = 853 + 7.5 T + 5.4 N

where T = effective transpiration in mm; N = total tonnage of N in 1000s of tons, roughtly equivalent to millions of kilograms. The details are given in Table XXXVI and illustrated in Fig. 17. The mean error is about 60 kg/ha, or $2\frac{1}{2}$%; the multiple correlation is 0.93. The single correlations are:

Yield on effective transpiration	0.79
Yield on total nitrogen	0.73
Effective transpiration on total nitrogen	0.26

The consequent partial correlations are:

Yield on effective transpiration	0.90
Yield on total nitrogen	0.88

The partial correlations from the previous analysis were:

Yield on effective transpiration	0.89
Yield on improvement factor	0.88.

This confirms that the greatest part of the improvement in hay yields was due to the increased use of nitrogen.

TABLE XXXVI

Verification of estimates of meadow hay yields

Year	Effective transpir. (mm)	Total N (10^6 kg)	Calculated yield (kg/ha)	Actual yield (kg/ha)
1939	196	9.7	2,375	2,485
1940	165	10.5	2,147	2,146
1941	188	11.7	2.326	2,422
1942	198	12.9	2,408	2,397
1943	203	14.6	2,454	2,472
1944	163	16.6	2,165	2,033
1945	213	20.4	2,560	2,610
1946	213	25.0	2,585	2,447
1947	188	26.7	2,407	2,372
1948	223	30.7	2,691	2,585
1949	191	36.2	2,480	2,497
1950	198	39.2	2,550	2,610
1951	213	45.4	2,685	2,610
1952	211	44.8	2,677	2,723
1953	211	55.2	2,733	2,861
1954	203	62.4	2,712	2,686
1955	224	69.4	2,908	2,886
1956	178	74.5	2,590	2,598

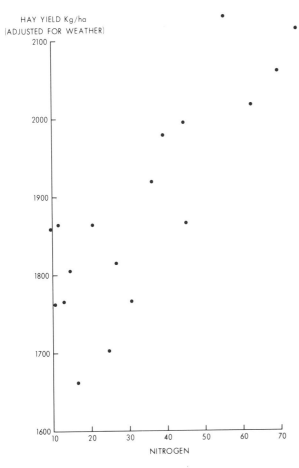

Fig. 17. National hay yields and use of nitrogen fertilizer (millions of kg).

The leaching factor

The decision as to how much nitrogen should be applied to crops depends a great deal on the leaching factor which is determined by the amount of excess winter rain. The estimation of this excess is a relatively simple process provided that the dates of the autumn start and spring finish of the running of the water in the drains or drainage outfalls is known. Assuming a negligible amount of surface run-off, the required answer is the total rainfall between these dates less the evaporation and transpiration during the same period. In winter the evaporative loss is very small, and errors in its estimation are not serious, such errors being very much of the same order of magnitude as the possible 5% error in the rainfall total. The final answer of excess rain for present purposes, namely the estimation of fertilizer requirements, is not needed to a high degree of accuracy. For this reason, if no other, it is practicable to rely on the calculated dates of return to field capacity for the start of the excess rain period; the end of the period is determined by the time when transpiration exceeds rainfall in the spring and a new season soil moisture deficit begins to build up (p. 47).

Amounts of excess winter rain can thus be found for any rainfall station for which estimates of potential transpiration are available. The simplest approach to an areal value is to take the arithmetical mean from a number of sample stations. The answer will not be absolutely correct, because of the inadequacy of the sampling, but the relative value from one year to the next is reliable as an aid to advice regarding lime and fertilizer requirements.

If an indicator value of this type is insufficient and a more accurate areal mean for a given year is required, then a more sophisticated method must be used. One such method involves the use of long-term averages.

Within an area of broadly similar potential transpiration and distribution of montly rainfall, there is a very high correlation between the average annual rainfall (R) and the average excess winter rain. As a result a series of regression equations can be calculated, such as those shown in Table XXXVII.

Averages of excess winter rain can therefore be calculated for any site for which the average annual rainfall is known or can be estimated. Furthermore, if the areal mean average annual rainfall is known for any region, preferably one with little variation in rainfall regime, then these formulae will provide the required regional average of excess winter rain.

In any one year the percentage of average of excess winter rain at a series of sample stations can then be used to estimate the areal value.

For example, the county of Cheshire in the West Midlands has an average excess winter rain of 320 mm. Two sample stations within the county have averages of 226 mm and 348 mm.

In the year in question the excess rain at these two stations was found to

TABLE XXXVII

Regression equations for excess winter rain

Area	Equation for excess winter rain* (mm)
England — North	$0.80\,R - 315$
— Midlands	$0.83\,R - 343$
— East	$0.82\,R - 343$
— Southeast	$0.77\,R - 320$
— Southwest	$0.82\,R - 345$
Wales	$0.90\,R - 409$

*R = average annual rainfall in millimetres.

be 136 mm (60% of average) and 228 mm (64% of average), respecitively. The estimate of the areal value for Cheshire would then be:

$$\frac{60 + 64}{2} = 62\% \text{ of } 320 = 200 \text{ mm (approx.)}$$

To guard against the possibility that these two samples were not representative, similar percentages for sample sites in counties surrounding Cheshire could be used. Taking four such extra samples, one to the north, one to the east, one to the south and one to the west, their percentages of average were found to be 57, 75, 74 and 66%, respectively. Using weighting factors of 1 for a sample within the county and 1/2 for samples outside, the result would be:

$$\frac{60 + 64 + \frac{1}{2}(57 + 75 + 74 + 66)}{4} = \frac{260}{4} = 65\% \text{ of } 320 = 208 \text{ mm}$$

Averages of excess winter rain and the variations about such averages can be a very important agro-climatological parameter.

MULCHES*

The use of a mulch, which is the name given to a layer of additional material placed on the surface of the soil around growing plants, is a common practice in many areas and for a wide variety of plants. The mulch can be of vegetable material such as straw or compost, or it can be manufactured, such as plastic. This alteration to the nature of the soil surface has a complex but significant series of effects on the microclimate of the top layers of the soil and the layers of air immediately above it.

The obvious aim of mulching is to increase crop yields, so that it might be thought that any experiment designed to test the efficiency of the mulch need only be interpreted in terms of yield with, and yield without, any form of mulch. This is far too simplistic a view, because to explain the results and to be able to apply the experience gained for future use, or for use elsewhere in different climatic conditions, it is necessary to understand exactly how the plant climate has been altered. An advantage gained by the use of a mulch in a certain set of growing conditions might turn out to be a disadvantage at another time of the year or in weather which was slightly, but significantly, different.

It is therefore advisable to take appropriate meteorological measurements

---

* The World Meteorological Organization, Geneva, are (in late 1974) about to publish an extensive review paper, prepared by J.W. Davies, on the subject of mulches in their *Technical Note* series.

during the course of any experiment. Before deciding on the nature and extent of such measurements, it is wise to reconsider the theoretical changes in crop climate which would be caused by a mulch.

Any layer over the soil will affect the flow of heat from air to soil and from soil to air; the temperature regimes in the soil and in the layers of air nearest the soil surface will therefore be changed. If the net effect is one of insulation, less heat will enter the soil by day and soil temperatures will consequently be lower, meanwhile air temperatures by day may be increased. At night, less heat will move from soil to air, thus increasing the frost risk in the air, but retaining heat in the soil. Surface films of plastic, on the other hand, can transmit radiation of certain wavelengths and may produce an increase of soil temperatures both by night and day. A black coloured mulch will absorb more radiative heat, but a white surface will reflect more heat back into the air around the plant.

Over a period of days or weeks, with widely varying weather conditions, the net effect of any mulch can be totally different from what at first thought would be expected, simply because the imagination cannot weigh up the plus and minus factors correctly. It is therefore necessary to take soil temperature measurements at various depths, several times per day. Air temperatures near the surface should also be measured, especially the minimum values at the upper surface of the soil cover.

The mulching layer will affect the water balance, but again the effect is by no means simple. A mulch will restrict evaporation from the soil surface, but it will also prevent light rainfall from entering the soil and root zone. It will tend to reduce weed growth, which will help to conserve soil moisture and soil nutrients. It will tend to prevent the capping or panning of the soil surface, and therefore facilitate the infiltration of water and the carbon dioxide exchange. The process of irrigation has to be modified, for it cannot be done in small amounts which would barely penetrate the mulch layer to the soil beneath and so provide little benefit to the crop. Larger amounts have to be added at longer intervals, and care must be taken in applying topdressings of fertilizers.

The measurement of soil moisture is always a difficult process, but attempts have to be made to quantify this factor, even if only in general terms. Soil sampling and oven-drying is possibly the most reliable method, but it certainly is the most time-consuming. Tensiometers or soil-moisture resistance blocks can be used if sampling errors are taken into account and absolute accuracy is not required. Neutron-scattering devices have become more widely used in recent years, but they are expensive and liable to error in the most important layers of soil closest to the surface.

Mulches may indeed prevent the entry of light rainfall into the soil, but they offer valuable protection against the adverse effects of heavy rain. Similarly they are an effective counter-measure against erosion, especially on sloping ground. It is very difficult to place a numerical value on these beneficial effects, but a series of observations on a bare soil surface before and after heavy rain can give some guide to the danger potential. During any mulching experiment it is essential to measure rainfall amounts and rate of rain recorders can give important additional information.

Measurements of rates of evaporation are rarely possible unless sophisticated micro-meteorological equipment is available. Some idea of the order of difference can be obtained by the use of lysimeters, but site-to-site variations in the readings provided by such instruments are too great to place undue reliance on the exact values of the results.

In regard to crop measurements, attention must be paid to the phenological stages of growth as well as to the final yields. In some cases, assessments can be made of crop quality, such as size, appearance, taste and constituents.

In the final published results of any experiment, it is important to include a full description of the mulch, its nature, its depth, time of application and changes in quality over the period of the observations. Exact details must also be given of the manner in which the meteorological measurements were made.

Even with extensive meteorological information, it can often be very difficult to explain the biological results in terms of the physical conditions. This, however, does not excuse the absence of any attempt to define the altered plant climate. It is only by consideration of the fully described conditions and results of many experiments that progress can be made to understand the complex situation. Failure to provide the full details means that little use can be made of the results by other experimenters.

It is always wise, whenever possible, to run a series of mulching experiments at several sites with different climates over the same period of time. With a restricted number of sites, it is often necessary to repeat the experiments over several years to obtain the necessary wide sample of growing conditions. It is dangerous to draw inferences from a restricted experiment, or to extrapolate the findings to other regions or other years.

SUGGESTIONS FOR FURTHER READING

Fertilizers

Smith, L.P., 1962. Meadow hay yields. *Outlook on Agric.*, 3: 219—224.

Work days

Smith, C.V., 1970. Weather and machinery work days. *Univ. Coll. Wales, Aberystwyth, Mem.*, 11: 17—26.
Smith, L.P., 1972. The effect of weather, drainage efficiency, and duration of spring cultivations on barley yields in England. *Outlook on Agric.*, 7 (2): 79—83.

Soil conditions

Smith, L.P., 1971. The significance of winter rainfall over farmland in England and Wales. *M.A.F.F. Tech. Bull.*, 24: 69 pp.

Smith, L.P., 1971. Assessment of the probable dates for return to soil moisture capacity in the autumn. *ADAS Q. Rev.*, 2: 71—75.

Smith, L.P. and Davis, J.H.R.H., 1972. Autumn cereal sowing and the date of return to field capacity in the soil. *ADAS Q. Rev.*, 6: 36—40.

Smith, L.P. and Croxall, H.E., 1973. Autumn weather and cereal growing in England. *ADAS Q. Rev.*, 8: 177—183.

Smith, L.P., Cochrane, J. and Bailey, V., 1973. Estimation of wheat and barley acreages in England by reference to the weather of the previous summer. *Meteorol. Mag., Lond.*, 102: 265—269.

CHAPTER 6

The Modes of Agricultural Meteorology — Techniques

> "Although this may seem a paradox, all exact science
> is dominated by the idea of approximation."
> BERTRAND RUSSELL

> "The best is the enemy of the good."
> VOLTAIRE

IRRIGATION

Many experiments have been carried out in numerous countries and climates in attempts to find out the value of irrigation, but the reports of such experiments which finally find their way into published papers rarely give adequate information about the soil-moisture regime. Some do not even include details of the natural rainfall, and the inclusion of data on potential transpiration is very rare indeed. As a result it is very difficult for the reader of such a paper to comprehend the circumstances under which the experiment was conducted and it is well nigh impossible for him to draw any major conclusions from the findings.

Irrigation aims at creating the best state of soil moisture relative to the crop and the crop stage, so that it is essential to make some attempt to estimate or measure the soil moisture conditions. Very few irrigation experiments fail to give some sort of response in terms of crop growth or development, but unless the details of the soil and air climate are given it is difficult to try to understand why such responses occurred. The efficient measurement of soil moisture is by no means easy or simple, but it should always be possible to make some attempt at drawing up a water-balance sheet and so get some idea of the moisture regime.

Rainfall measurements, on a daily basis, are mandatory, but also some estimate of the potential transpiration should be made, even if it is only a parallel measurement of a pan or tank evaporimeter. If instruments of this type are used, it must be remembered that errors of up to 20% are possible due to almost imperceptible variations in exposure, and it is always wise to use them in duplicate. The use of lysimeters, superficially a very attractive method of experimentation, can lead to even greater errors. In general, it must be stressed that any parameter which is obtained by the subtraction of one measurement from another, such as change in soil moisture content, can be subject to far greater errors than those of the individual measurements; errors of 10%, quite common in field measurements, can lead to errors of 100% if a process of subtraction is involved.

Irrigation experiments are generally designed to find out the water requirements of a crop and to identify the stages in its development when it is highly sensitive to moisture stress. If the water-sensitive stages are found, and if the water requirements can be expressed in terms of the current weather, then the results of the experiments can be used as the basis of a meteorological information service to irrigation practice on a field scale.

An irrigation trial can, in fact, be used as a check on transpiration estimates. It is assumed that if too much water is applied (the soil being restored to beyond field capacity), then a crop yield will be either reduced or at least show no increase over the maximum. If the experiment shows that the treatment giving most water shows a yield depression, then it can be inferred that the transpiration estimates may be too high. Conversely, if a treatment deliberately designed to give too much water still shows a yield improvement, then it is possible that the estimates are too low.

Care must be taken in plot trials. Unless adequate (large) guard rings are used, the microclimate of one plot will "flow over" into adjacent plots and cause utter confusion of the microclimatic vertical profiles of temperature and humidity, so that plot transpirations will not be what they seem. It follows that the usual randomized plot lay-out, beloved of statisticians, may by no means be the best arrangement, and a better system may be a series of parallel strips. If plot 1 is the control, receiving only the natural rainfall, and if the other treatments are in plots 2, 3, 4, and 5 with increasing water amounts, then a series of parallel plots athwart the prevailing wind, numbered 1-2-3-4-5-5-4-3-2-1-, will minimize the inter-plot effect.

It is not easy to devise a method where one plot does not get any rainfall at all. Any form of cover will affect the growing climate around the plants and the consequent growth. A moveable cover, used only during rain, presents difficult management problems.

Field trials are in many ways preferable to plot trials, but are impossible to carry out with limited resources. Small-scale trials can sometimes give misleading results, but irrigation is so important a means of improving crop production, that work in this aspect of agro-meteorology is fully justified. In capable hands it can give results of the highest significance and great potential use.

SHELTER

Shelter from the wind for growing crops comes in various shapes and sizes, ranging from a tall tree shelter-belt to a small hedge and including various forms of non-living barriers such as walls, fences, netting and plastic materials. Countless experiments have demonstrated that crops benefit from the proximity of shelter provided that they are not shaded from sunshine and there is no serious competition for soil moisture and nutrients. Measurements have also been made of the weather conditions in the lee of shelter, showing the resultant changes in such factors as wind speed, air and soil temperatures, humidities, evaporation and transpiration.

The difficulty arises when attempts are made to explain the gain in yields of crops in terms of the observed changes in the weather factors. The changes in temperatures are very small, and so are the changes in evaporation measurements from tanks or pans. Evaporation from evaporimeters of the Piche type show marked changes, but such instruments are very sensitive to air movement and are probably largely acting as anemometers.

Bearing such small changes in plant climate in mind, it seems impossible not to conclude that it is the wind flow through and over the crop which has a dominant effect on the photosynthetic process, so that it seems advisable for shelter experiments to concentrate on wind measurements and crop development and yield.

Such experiments are much easier to plan and carry out when they involve the smaller type of windbreak, especially when non-living materials are under review. Different designs and alignments can be compared, and close attention must be paid to the degree of permeability of the materials.

Experiments with large tree shelter-belts take much longer to carry out, and if it is wished to measure the changes in wind regime during the growth of a newly planted belt, several years will be needed. Such a belt was planted on an experimental farm in north England, and two cup-counter anemometers were installed from the start, one 25 m to the east, the other some 300 m to the east of the belt of trees, which initially was less than 1 m in height. The percentage decrease in wind at the site nearest the shelter-belt is shown in Table XXXVIII. This series of indices of relative shelter show how the effect built up through the years as the trees grew in height, and also indicates the effect of reduced winter shelter when the deciduous trees shed their leaves.

TABLE XXXVIII

Progressive effectiveness of a growing shelter-belt

Summer	Percentage decrease in wind	Winter	Percentage decrease in wind
1952 (June)	8	1953 (January)	4
1953 (July)	10	1954 (January)	3
1954 (July)	11	1954 (December)	11
1955 (May)	11	1956 (January)	14
1956 (August)	24	1957 (January)	29
1957 (July)	26	1958 (January)	21
1958 (July)	46		

Note: All percentages refer only to days when the wind was blowing through the shelter-belt towards the two anemometers.

The horizontal extent of the shelter depends on the height of the shelter itself, so that it is customary to measure distances downwind in terms of multiples of the barrier height, enabling site to site comparisons to be made. During the period of any experiment, care must be taken to obtain complete records of wind direction from a freely exposed site, otherwise analysis of shelter effects becomes difficult if not meaningless.

Unless the terrain surrounding the experimental site is very flat and level, the effects of topographical shelter must be remembered. This effect of land conformation can form the subject of a series of experiments in itself, and surprisingly large effects on wind speed can be observed within a local climate. Again wind direction plays a very important part, and the experiments must run long enough to include a fair sample of possible types of weather. When it is needed to obtain a detailed analysis of the possible wind fields, it is often useful to supplement a fixed network of wind observations by several periods of intense measurement, employing several observers with mobile or hand-held anemometers.

HOUSING

It is reasonable to assume that the traditional buildings used for livestock housing have, by a slow system of trial and error over the centuries, reached a reasonably efficient standard of design and construction. Sensible reasons for a choice of site were also not unknown to our predecessors, and if a farmhouse has stood in the same place for many years there is every reason to believe that the choice was a good one.

Recent changes in building practice and so-called planning, especially in the "developed" countries of the world, have not all been for the good. New and larger powered farm machinery, new building materials and techniques, plus the tendency towards larger farms and bigger livestock units have introduced a new set of problems, many of which have not even been recognized, let alone solved.

It is probably true to say that the interior climate of modern buildings is often worse than that inside the old-fashioned types they have replaced. Errors have been made and money has been wasted simply because there was insufficient knowledge available of the basic physical laws governing the creation of a suitable living climate. An architect's dream can easily become the inhabitant's nightmare. Furthermore, if the building design and structure facilitate the incidence and spread of disease, what was merely discomfort becomes lethal in effect.

The problems are made more difficult by the fact that the ideal climate is probably unattainable in practice. If undue attention is paid to one environmental factor which is thought to be highly desirable, it often happens that a deterioration occurs in another overlooked factor. Slow historical development had reached a workable compromise solution; the new instant design was more liable to serious mistakes.

Once the desired climate is known, and this information is by no means always available or decided beyond dispute, the problem becomes one of site selection, structural design, and formulation of the correct operating procedure (heating, ventilation, etc.). Long-period experience having been discarded, such problems have to be solved on a sound scientific basis to have any chance of success. Not only has the external climate to be understood, but also the effect of the external on the internal environmental conditions. This understanding involves consideration of the heat exchange, the moisture exchange and the consequent temperature and humidity regime. The true facts of the ventilation must be known, not the wishful thinking or the false reasoning. The whole aerobiological problem of the air-borne transport of pathogens must be given full consideration, for an otherwise excellent building could quite easily become an efficient disease trap.

While it is possible by a certain amount of experienced thought to avoid the major mistakes of bad siting, incorrect structure and inefficient control, a degree of assessment of environmental conditions by the use of instruments is generally necessary to solve any particular housing problem. Such measurements have to be taken while the building is in use, because an unoccupied space has quite a different climate from one filled with animals. This creates difficulties in the siting and exposure of instruments; an ideal site may invite damage by inquisitive animals.

Although the maximum use can be made of conventional instruments, if only for the fact that they are readily available, it may be necessary to design special equipment. Certainly measurement of ventilation or air change demand techniques which are new to most meteorologists. The *W.M.O. Technical Note* No. 122, "Some environmental problems of livestock housing", contains useful advice and guidance in regard to these problems. One further word of caution may be added; if use is made of hair hygrometers, great difficulty will be experienced in trying to keep them free from dust, and their readings will always be suspect.

The agro-meteorologist, especially one who is associated with an agricultural advisory or extension service, may often be called upon to act as a consultant to help to solve problems which have arisen in a given set of buildings. Whatever mistakes are involved, they have already been made, and the purpose of the investigation is firstly to identify the errors and then to suggest modifications in building structure or management practice to minimise or eliminate them. It is too late to correct errors in siting. For example, a calf-rearing unit which has been placed in a frost-hollow, simply because it appeared to be a nice sheltered situation, will present problems concerning the infiltration of cold night air. It is hardly practicable to adopt the best solution of moving the whole expensive unit.

During the course of such investigations a sense of important values is built up. Techniques of measurement and analysis are acquired which cannot be learnt without hard experience. Gradually it becomes possible to specify the causes of major potential error, and it is then that design standards can be improved. Once this is done, the agro-meteorological advice can be made available at the planning stage, so that previous mistakes made elsewhere will not be repeated in any new enterprise.

The solving of an existing problem is a valuable aid to production at the farm in question, but this is equivalent to giving first aid after an accident. The real permanent help is to prevent the occurrence of such accidents.

The farmer may not be aware of the climatic implications of a proposed site, and neither may the planning authority which gives him permission to build. The builder or designer may not understand the type of internal climate he is creating, especially if he is using new and relatively untried materials or designs. The farmer or his staff may not know how to manage such buildings and husbandry methods may be employed which were suited to old-style housing but which are quite inappropriate in the new context.

A case in point is the introduction of excessive ventilation, possibly desirable in itself, but resulting in a much larger diurnal variation in temperature which can be dangerous for young animals. Conversely, harmful draughts may be removed, only to find that pockets of stagnant air are created. It is surprising to the meteorologist that many people do not realize that a reduction in temperature almost invariably results in a rise of relative humidity.

The agro-meteorologist, in the initial stages, may feel that he knows very little about the complex problems; he may rest assured that other people probably know considerably less.

STORAGE

The physical conditions of the storage of fodder for animals, or of food for humans, is another problem which must not be forgotten by agricultural meteorologists. The losses in store, on a world scale, have been estimated to amount to about 20%.

The basic problem is essentially the same as that concerning the housing of animals. It is first necessary to determine the storage conditions which are desirable, and which will not cause losses due to the onset of pests or disease. The second part of the problem is to decide how these conditions can be provided, involving the best selection of the site, design of the structure holding the stored materials, and management thereof.

Whereas the placing of a building designed for animals in a frost hollow or donor area for cold katabatic winds on radiation nights may be a serious disadvantage, the same site may be exactly the correct place for, say, a potato store which has to be maintained at a relatively low temperature to avoid sprouting. Areas which are susceptible to long periods of high humidity are undesirable for the storage of grain or fodder such as hay. The correct selection of a site is most important, because the best of management cannot overcome the liabilities of a bad local climate without a large expenditure of time and money.

Temporary, or semi-permanent storage may be effected by the use of clamps, field-side heaps of sugarbeet, potatoes, or grass silage covered by earth, straw, or plastic sheeting. The climate under such covers may not be adequate for the prevention of loss or deterioration of the stored material. It is relatively easy to measure the internal temperature conditions in such clamps, and to find its reaction to changes in the outside weather. Although local farming practice is likely to have reached a fair degree of competence in this matter, there is often plenty of scope for simple improvements.

Research into most storage problems, however, will be a more complicated affair, and will need the active cooperation of other scientists such as entomologists and pathologists.

GLASS AND PLASTIC COVER

The meteorological aspects of the siting, design and operation of heated structures of glass or plastic are so complicated that they deserve a book in themselves. Commercial practice involves efforts to control light, temperature, heat, humidity and carbon dioxide in an attempt to produce the best growing climate for the crops. To do this, measurements are made of the internal climate but it is not always realized that great care must be taken to get a representative value. There exists a very appreciable variation in the physical conditions in the various parts of the glasshouse (in Britain, the word greenhouse is used only to refer to small glass structures for amateur gardening). Attention must be paid to the sampling procedures, as well as to the problems of instrument exposure.

Before starting any experimental work on the effect of glass cover, it is as well to consider what exactly are the physical processes involved. The most common explanation of the effect of glass is that it transmits short-wave radiation in, and reduces the long-wave radiation out. While correct as far as it goes, this is far too simple an explanation for the heating effect, and it is often forgotten that there is considerable conduction of heat through the glass, so that on a cold night the temperature of the inside glass surface is very little different from that on the outside. Glass cover retards heat loss but cannot entirely prevent it. Furthermore no glass structure is air-tight and wind-proof, so that the advective loss of heat is not negligible. Nevertheless it is probably true to say that the greatest single effect of glass is the cutting down of the wind speed, so that there is a greater increase of soil temperature by day, and a consequent greater return of heat from the soil to the air by night.

A transparent cover does transmit most of the incoming radiation, but it is bound to decrease the light especially when the cover is dirty; this may be very significant during the winter months when the natural light intensity is low. The transmission properties of glass and the various forms of plastic are not the same, and each type of material needs separate consideration and experiment.

Factors to be taken into consideration when selecting the site for a glass-house include the external regimes of radiation, light, temperature and wind. The solution is rarely simple, and almost invariably a compromise has to be reached. For example, although the sea coast generally provides the best chance of a good sunshine record, it also is most susceptible to gale force winds and contamination of the glass by salt spray. An inland site will have less wind problems but be more liable to low night minima.

Taking into account the implications of the rapidly dwindling sources of supplementary heating fuel, it becomes more and more important to make the best use of heated structures. On a strictly logistic basis, far more energy is used than is produced in the form of an eatable crop, and it may not be long before it is no longer economic to use fuel in this way. On the other hand, unheated glass or plastic structures may face a far more promising future.

Experimental work on the meteorological conditions inside large glass-houses demands a heavy expenditure of money and materials. It is therefore normally only possible to do such work at special research stations with adequate resources of man-power and support facilities. A great deal of useful work on a smaller scale can nevertheless be done by investigating the conditions under low glass or plastic covers in the form of cloches or frames.

For example, an experiment was carried out at Bristol in 1960, using seven minimum thermometers placed inside the centre cloche of a row of nine over a short grass cover. Four of these thermometers were placed on the glass sides and sloping roofs, touching the glass; three others were placed in a vertical line in the centre of the cloche. Readings from these instruments were taken on over 60 nights and were compared with the temperatures observed by a conventionally exposed grass minimum thermometer in the open.

On the nights with the greatest difference in temperature, inside—outside, the four thermometers on the inside glass surface showed values some $2°-3°$ F higher than that of the outside grass minimum. As regards the values in the centre of the cloche, the thermometer at the lowest height showed a gain of some $5°$ F, and those above it about $4°$ F. These results confirmed the theory that the gain in night minimum under glass is greatest away from the sides of the structure and nearest to the soil which has been warmed during the day. On cloudy or windy nights, when there was little or no radiation cooling, there was little difference between the inside and outside minimum temperatures.

Further analysis showed that on radiation nights, which were defined as those with no precipitation, a wind less than 6 knots and a sky less than half covered with cloud, the extent of the gain in minimum temperature depended on the amount of sunshine experienced the previous day and on the strength of the temperature inversion in the lowest 2 m of air. This last factor is conveniently measured by the difference between the screen minimum and grass minimum temperatures, which was found to vary from some 3° to 10° F on such radiation nights.

A similar later experiment investigated the night conditions under Dutch lights, a type of unheated glass frame about 2 m square, used in pairs to effect a sloping roof-like structure with the ridge 50 cm high and the sides, which are boarded with wood or brick, 8 cm high; the total span of covered soil is usually about 3 m. One set of such lights was sprinkeld with water whenever the outside temperature fell below 1°C, and a series of observations over the December to March period showed: (a) the gain in night minimum surface temperature under the lights on radiation nights averaged 5°C, and on occasion exceeded 7°C; (b) the extra gain produced by water sprinkling averaged 2.5°C, and on the coldest nights approached twice this value.

Despite the fact that, in the open, frost prevailed for up to 15 h, on no occasion was a temperature below freezing point experienced inside the sprinkled frame (see Fig. 18).

Some of the incoming radiation of the following day has to be used to melt the ice which forms on the glass surface, but even so, the maximum

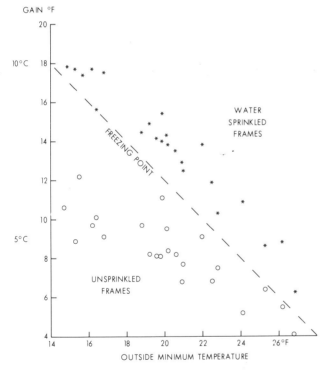

Fig. 18. Increases of minimum temperature caused by the continuous sprinkling of water over glass frames on frosty nights.

temperatures inside the frame do not seem to be effectively lowered unless they exceed 15°C. There thus seems little doubt but that the combination of good glass cover plus water protection affords an excellent defence against winter frost conditions.

The avoidance of frost is only one way of measuring the beneficial effects of crop cover. A further important measure of its benefit is the way in which it extends the effective growing season. The main disadvantage of such cover is the necessity to compensate for the absence of incoming rainfall, so that for most crops, some form of irrigation is needed. The only exceptions are those quick-growing crops such as lettuce which can complete their growth using only the moisture stored in the soil, but in this case it is essential that the soil is at capacity before the crop is covered.

The problem of water requirements under glass can be the subject of a whole new series of field investigations in agro-meteorology. Experience has shown that the calculated values of potential transpiration prepared for outdoor crops can still have a reliable degree of validity, despite the fact that the inside climate has been considerably altered. At very least such calculations can be used as a reference point for devising the experimental treatments.

SHADE

In the so-called temperate climatic zones there are few occasions when growing crops need shade from the sun. On very hot summer days, grazing animals may have the need for shade, and their food intake may be reduced, but as their energy output is also reduced, the effects are rarely very serious. When glass or plastic crop cover is used in horticulture, a need for summer shading is often met by coating the cover with emulsion paint or similar easily removable material.

In lower latitudes, with longer hours of more dependable sunshine, shade becomes a far more important problem. Very little can be done to protect most tropical crops, except those such as coffee for which it is common practice to grow within the shade of larger trees or bushes. Traditionally the tea crop is said to require shade, but the density of the trees seen in tea plantations is far too low to produce any effective result as far as shade is concerned, although they may produce some diminution of wind. There seems to be little scientific justification for this practice.

Shade for animals in tropical climates is a far more serious problem, the solution of which is made easier by the fact that the animals are mobile and can seek any available shade when the conditions so demand. When the sun is at a high elevation, almost directly overhead, a man whose body is roughly equivalent to a vertical cylinder intercepts the minimum of direct insolation, far less for example, than when the sun's rays are at an angle to the vertical and influence the side of his body. The erect man is therefore affected least when the sun is overhead at midday, although the air temperature may prevent him working. A farm animal more closely resembles a horizontal cylinder and so, no matter what the elevation of the sun, it presents an approximately constant surface for the interception of the sun's rays.

This radiation overload on animals has a dual effect; directly, it brings about a stress on the heat balance, causing it to seek means of reducing its skin temperature by standing in shade, wallowing in mud, or by panting to increase evaporative loss of heat; indirectly it curtails the length of time spent in grazing and consuming food; the combined result is that the animals do not thrive. Wild animals, in such conditions, have learnt to adjust their behaviour, but domestic animals, especially introduced breeds, can experience great difficulty, although acclimatisation does take place and new behavioural routines are learnt.

In high radiation conditions, good animal husbandry aims to provide the herds and flocks with access to shade, either natural or artificial. The duty of the agricultural meteorologist is to help the animal physiologist define the environmental need for shade and the extent to which it will be used, taking into account the possibilities of supplementary feeding such as "zero grazing".

The basic meteorological observation is thus some measure of radiation or sunshine hours which can be interpreted into an assessment of heat load on the animal. Observations of animal behaviour in regard to the use of shade, when available, can be analysed with respect to the weather factor as a first step towards defining the threshold conditions of excess heat. The intensity of incoming radiation may be the dominant factor, but it may not be the only one. Wind strength and humidity should also be considered, and there may be an indirect weather effect concerning the availability of fodder. An animal may have to spend longer hours in grazing if food supplies are sparse and the range is in poor condition; it may have to cover longer distances to get the food it needs. It is probable, therefore, that two threshold conditions need to be defined, one concerning the occurrence of adverse heat stress, and the other defining the minimum desirable hours of free grazing. Experience in tropical climates has suggested that "zero grazing", bringing the food to the animal and keeping it within a shaded area, may be the most efficient way of solving the double problem.

Once the results of a field trial have been analysed and the critical weather conditions and grazing conditions established, it becomes a problem of agroclimatology to put the results into planned practice.

There are times and places where shade can have considerable adverse effects. This can occur in higher latitudes with a weaker radiation climate and a lower elevation of the sun, especially in the short-day winter months. There is now a deficiency of incoming heat for plants and animals, and long shadows are cast by buildings, trees, or ranges of nearby hills producing significant changes in local and micro-climates.

Using astronomical or navigational tables, it is not difficult to calculate the potential loss of sunshine due to the shadow of any object, it being merely an exercise in three-dimensional solid geometry based on the altitude and azimuth of the sun and the dimensions of the obstacle.

It is less easy to interpret the effect of shadows, whether in regard to the duration of frost on a plant or the lack of ultra-violet light on an animal. Nevertheless, local problems may arise in which the shadow effect is a key factor. For example, a case comes to mind of severe scouring in a flock of sheep which were being given supplementary feed of swedes or turnips during the winter. The roots were laid out in the field close to a high tree shelter-belt, where the sheep were keeping together out of the cold winds. The shadow cast by the trees prevented a rapid daytime thaw of such food after an overnight frost. The sheep could not digest the frozen roots and scouring resulted. Once the food was provided outside the shaded zone, these symptoms disappeared.

Topographical shadows can have a recognizable effect on the earlyness of spring growth, on the phenological dates of crop stages of development, and on animal welfare. These effects are complicated by other factors such as wind regime. Grazing conditions may be better on the sunnier side of the hill and yet a hill flock of sheep may seek shelter from a southwesterly wind on the shaded side where the grass is growing less freely. A plant may grow better in a partially shaded site, merely because it is sheltered from the wind. It is rarely advisable to examine one factor in isolation when searching for agro-meteorological relationships.

TRANSPORT

The transport of food from market to consumer may be thought to lie within the responsibilities of industrial meteorology, but transport from farm to market can still be thought to be part of agricultural meteorology.

Produce from the land, vegetable or animal, has to be sold in good condition before it contributes to a farmer's income or to the gross national product. Deterioration can take place during transit for reasons which are often, in part, meteorological. With a little knowledge and a little foresight, many such losses can be avoided.

For example, soft fruit can rapidly lose its fresh quality in a short time after picking especially during hot weather. If the fruit is packed and loaded on to a vehicle which is then left on a concrete hard-standing in full sunshine on a warm summer day, the resulting deterioration can prejudice the sale of the fruit. A lorry or truck in transit will not bring about quite the same effect, because the speed of movement will produce a ventilation effect and keep the fruit temperature close to that of the air, but even so, travelling through the heat of the day is not advisable, unless some refrigeration is available.

Cold weather can also bring problems; potatoes, whether for ware, or more especially for seed, cannot be moved by road or rail during frosty weather without risk of losses. Some vegetables, such as celery or sprouts are far more resistant to frost.

Rain at picking or harvesting can also cause losses, especially for fruit, but the only way to avoid this is to stop picking. Many problems of potential losses in transit demand access to accurate weather forecasts because a change of timing is the only way to avoid the adverse conditions.

Animal losses during transport can be serious and the exact relationship between such losses and the weather deserves careful analysis. For example, a load of pigs being taken by road transport to the slaughter-house are sometimes found to include a number which are dead on arrival; others may subsequently die while being kept in pens before actual slaughter; others may be in such a poor condition that they have to be killed to avoid further suffering. When slaughtered these pigs frequently disclose pale watery meat which is virtually inedible.

The veterinary explanation of such deaths is now called an "acute stress syndrome", but other names such as "acute cardiac death", "muscle degeneration" and "malignant hyperthermia" are used. The important meteorological factor is that temperature during transport appears to have a strong controlling effect on the extent of the losses.

The possibility that this could be true is immediately suggested by the plot of the losses per 1,000 on a monthly basis. Fig. 19 shows the mean monthly losses and the mean monthly daytime temperature over a period of 12 years, for one abattoir. Examination of these graphs and simple numerical correlations furthermore indicate that temperature is likely to be more significant than radiation (because the June losses are lower than July), and also that the temperature—deaths relationship may not be the same in the January—July period and the August—December period (because August losses show a pronounced fall from those of July). The daytime temperature, meaned over hours of daylight, was thought to be most appropriate to the likely times of transit although a mean taken over the working day would probably have shown little difference in pattern.

The linear correlations were found to be:

	Jan.—July	Aug.—Dec.
Mean daytime temp. on mean deaths on arrival	0.97	0.99
Mean daytime temp. on mean deaths in pens	0.95	0.98
Mean daytime temp. on total deaths	0.96	0.99

Meaned values of a period as long as 12 years can, however, hide important features. The data was therefore re-analysed in terms of the mean daytime temperatures of individual months, which were grouped together in overlapping ranges of 3°C. The results are shown in Table XXXIX. The apparent

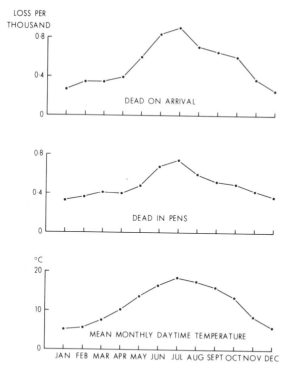

Fig. 19. Seasonal variations of deaths of pigs in transit and mean daytime temperatures.

linear relation over the whole range of temperature has now disappeared and has been replaced by a two-part structure incorporating a slow increase with temperature up to 9°C and a rapid increase thereafter, this latter increase being most marked in the late spring and early summer period.

The correlations are:

	Jan.—July	Aug.—Dec
Below 9°C	0.91	0.96
Above 9°C	0.99	0.93

To proceed to greater detail, it now remained to examine the data on a daily basis. First, however, it was necessary to make sure that temperature was in fact the dominant meteorological factor. For example, it might be thought that atmospheric humidity would have some effect on the stress conditions. The dewpoint at mid-day was selected as a convenient humidity parameter. There is some doubt as to the exact day on which the pigs were reported as "dead in pens", so only the data for "dead on arrival" were considered. For each of the 1°-temperature categories, the days were divided into those with deaths on arrival and those without, and the mean dewpoints found within each class. It was found that there was no consistent difference

TABLE XXXIX

Deaths of pigs in transit and ambient temperature

Temp. range (°C)	Deaths per 1,000 pigs carried	
	Jan.—July	Aug.—Dec.
Below 4.9	0.63	0.53
3— 5.9	0.63	0.52
4— 6.9	0.61	0.61
5— 7.9	0.66	0.69
6— 8.9	0.69	0.76
7— 9.9	0.73	0.74
8—10.9	0.76	0.82
9—11.9	0.81	0.81
10—12.9	0.92	1.05
11—13.9	1.01	1.10
12—14.9	1.06	1.12
13—15.9	1.21	1.12
14—16.9	1.36	1.13
15—17.9	1.50	1.20
16—18.9	1.55	1.24
17—19.9	1.66	

throughout the 1—19°C range. In the months January to July, for the 19 temperature categories, in 7 cases the mean dewpoint was higher in days with deaths, in 10 cases it was higher in days without deaths; in 2 cases the mean dewpoints were identical. In the remaining August—December months, the division was even more equal, 9 means were higher with deaths, 9 higher without deaths, and 1 identical.

The mean of dewpoints at a given temperature thus offers no evidence that humidity is a factor. This can be confirmed by examining the death rate for each dewpoint within a 3° temperature range. Within such a range, if humidity has any effect, the death rate should increase with increasing dewpoint. Eight such (overlapping) 3° ranges were considered from 15°—17°C to 22°—24°C. In no case was there an increase of deaths per 1,000 with rise in dewpoint. In fact, at the lower temperature ranges, a suggestion of a negative correlation was found, with higher losses at the lower dewpoints. In the four highest temperature ranges, the correlation coefficients of death rate on dewpoint were +0.19, —0.10, +0.23 and —0.41, which suggests that humidity is not significant. The explanation is probably that the pigs create their own humidity climate within the transport and that external humidity is unimportant.

Although direct sunshine may have little effect on the internal temperature of a moving well-ventilated vehicle, it is well known that such temperatures can rise quickly when the transport is at a standstill. The mean sun-

TABLE XL

Correlation coefficients; pig deaths and sunshine

Temp. range (°C)	Death-sunshine correlation
15—17	+0.22
16—18	+0.43
17—19	+0.59
18—20	+0.85
19—21	+0.50
20—22	+0.30
21—23	—0.10
22—24	+0.63

shine hours for days with deaths, and for days without deaths in transit were therefore found for each of the 19 temperature categories, as in the humidity analysis. The January—July results were 8—8—3, the August—December, 9—8—2, which again is a nil finding. However, in the higher temperature ranges some suggestion of a sunshine effect was discovered. Correlations of mean death rates with daily sunshine hours are shown in Table XL.

A more sophisticated analysis was then carried out, using the number of pigs transported each day as weighting factors. High correlations were found in the 17°—21°C range, as seen in Table XLI. This strongly suggests that within a fairly narrow temperature range (17°—21°C), the sunshine on the day of transit has some effect but as no simple way was available for incorporating this factor, the complete daily analysis was carried on using daytime temperature only.

It was first necessary to come to some conclusion as to which day the pigs reported as "dead in pens" actually travelled. Two methods were chosen, one involving the minimum adjustment wherein those dead in pens were allocated to the previous day when no pigs were carried on the day in

TABLE XLI

High temperature, sunshine and pig deaths in transit

	Weighted sunshine	Pigs transported	Death rate
18—20°C temp-erature range	1.41	63,253	0.82
	4.33	69,262	1.07
	7.07	53,087	1.11
	10.28	52,316	1.22
	12.75	74,993	1.51

Sunshine—death rate correlation of 0.96.

question (for example, on a Sunday); the other method involved a maximum sensible adjustment taking into account the numbers carried on the day in question and on the 1, 2 or 3, preceding days. In the event, the results obtained were very similar, so that clearly no great error has been introduced.

Furthermore, it could not be assumed that all pigs travelled during the time of day pertaining to the mean daytime temperature. Some would have travelled during the heat of the day, and some during cooler periods of early

TABLE XLII

Daytime temperatures, season and death rates of pigs in transit

Mean daytime temp. (°C)	Jan.–July mean deaths per 1,000	Aug.–Dec. mean deaths per 1,000
− 3.5	0.42	
− 2.5	0.55	
− 1.5	0.57	0.62
− 0.5	0.51	0.39
+ 0.5	0.55	0.41
1.5	0.60	0.50
2.5	0.64	0.51
3.5	0.67	0.53
4.5	0.64	0.59
5.5	0.63	0.64
6.5	0.62	0.65
7.5	0.62	0.70
8.5	0.63	0.76
9.5	0.68	0.80
10.5	0.80	0.83
11.5	0.88	0.81
12.5	0.92	0.84
13.5	1.00	0.92
14.5	1.07	0.96
15.5	1.17	1.08

Above this temperature, the Aug.—Dec. death rates fall appreciably below those of the Jan.—July period:

16.5	1.35	1.22
17.5	1.53	1.33
18.5	1.71	1.47
19.5	1.95	1.67
20.5	2.27	1.93
21.5	2.48	2.26
22.5	2.82	2.72
23.5	4.28	3.47
24.5	6.55	
27.5	20.29	

morning or late afternoon. To find the death rates of each degree of temperature, a weighting system of a 1/2, 1 and 1/2 was adopted covering a 3° range of temperature. This produced a very smooth curve so that the more complicated weighting of 1/3, 2/3, 1, 2/3 and 1/3 which might be equally justifiable was not necessary.

It was now found that the temperature—death rate relationship was no longer linear or bi-linear, but hyperbolic, the nearest fits being:

January—July   deaths per 1000 = $\dfrac{0.12X + 14.4}{28.2 - X}$

August—December   deaths per 1000 = $\dfrac{0.046X + 12.7}{27.5 - X}$

where X is the mean daytime temperature.

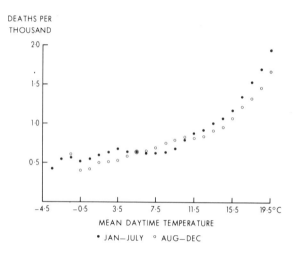

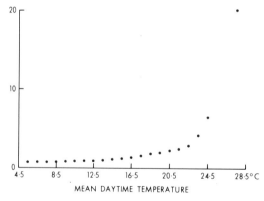

Fig. 20. Death rates of pigs in transit and mean daytime temperatures.

These formulae of course imply an infinite death rate at temperatures of 28.2° and 27.5°C, but this can be taken to be an exaggeration (see Table XLII). Insufficient days occurred within the 12 years to give meaningful death rates for 25.5° and 26.5°C but the dramatic increase at very high temperatures is fully substantiated (see Fig. 20).

It will be noticed that, except on very hot days, the death rate is so low as to be negligible, if humane reasons are ignored (a point of view that will not endear itself to any animal protection organization). In other countries, with other breeds of pigs, far more sensitive to stress, the losses can be very much greater. As the most sensitive pigs are often those which make the largest gains in live weight or attain the highest food conversion factors, this source of lost production may become more important in future years.

The above example of the analysis of an influence of weather shows how an initial idea can be developed and modified in the course of an investigation. It must not be assumed, however, that the results hold good for any other type of animal, or, as indicated above, any other breeds of pig.

Losses are experienced during the transit of young calves, but the seasonal pattern is completely different and it is obvious that high day temperatures are not the dominant factor.

Each problem must be treated on its own merits, and the results of one investigation cannot be assumed to hold true under changed circumstances. Although the results quoted above are concerned with one slaughter-house only, it is known that results from another slaughter-house for pigs, over a shorter period, fit a very similar pattern. Once confirmation of this type is available, planning to avoid transit losses can proceed with confidence.

WEATHER FORECASTS

As far as weather forecasts are concerned, the agricultural meteorologist has two main responsibilities. He must inform the weather forecaster of the farmer's requirements and he must explain to the farmer how to make the most use of the forecasts that are available.

Weather forecasting has provided a useful service to transport by land and sea. It might be thought that it could prove equally helpful to the similarly weather-sensitive operations in agriculture. There are, however, two cardinal differences. Firstly, an airline pilot is individually briefed with a specialized forecast, and very few farmers have access to an equivalent service, it being impossible to speak personally, on a question and answer basis, to a member of a meteorological forecast organization, except in relatively few telephone consultations. Secondly, it is most important to realize that an aviation or shipping concern has a choice of action. An aeroplane flight can be delayed or cancelled, it can be altered in regard to flight-plan in track or altitude, if the destination has unsuitable weather, landings can be made elsewhere. A farmer has few such choices; if he has decided on a particular plan of farming he has to go through with his plan right to the end. He cannot change his crop or move his fields, and his radius of manoeuvre is generally confined to a change in tempo, in that he can hurry up his operations to complete his work before the arrival of unsuitable weather, or he can wait a while for better conditions to ensue. Even so, he cannot wait forever; a crop ready for harvest has to be brought in somehow, whatever the weather. An aircraft finding its proposed destination shrouded in thick fog can fly elsewhere and land in safety. A ripe crop in pouring rain cannot be moved to a dry area and harvested. A ship can change course to avoid a storm, an aeroplane can fly round a violent thunderstorm, but an arable farmer receiving a forecast of gales or hailstorms can only sit and await the worst the elements can do to him. The livestock farmer is slightly better off because he can move his flocks or herds, but even then only within a limited range.

The requirements of agriculture

In order to plan efficiently the day's work on a farm, knowledge of the weather of the next 12—24 h is required to a very high degree of accuracy in regard to time, place and intensity. Temperature, humidity, wind strength and direction, rainfall duration and intensity can each be critical in certain operations. The desirable degree of precision, for example in regard to local frost intensity or occurrence of rain showers, is often impossible to attain with present forecasting standards. To complicate matters still further, the requirements vary with time of year, type of farm, type of soil and type of crop or animal. Thus a completely different state of affairs exists as compared to the relative uniformity of transport requirements.

The follow-up weather, usually referred to as the "outlook" period is also important. It is useless for a farmer to carry out a certain operation, for example the cutting of grass for hay, if tomorrow's weather is going to ruin the whole process. Knowledge of settled spells of fine weather is thus of the greatest importance especially in harvest time. On the other hand, after sowing, periods of intermittent rain or showers are equally desirable.

As mentioned earlier, the need to step up the speed of operations or the advisability of waiting a few days, is one which can be most helped by weather outlooks covering a period of days. A forecast for a week ahead must, as far as scientifically possible, indicate if the weather is getting wetter or drier, warmer or colder.

Forecasts for periods longer than a week are of limited use, for the tolerable change in timing is rarely more than a few days. It is true that crops may have to be sown late or harvest may be delayed by weeks or even months, but losses are then inevitable and these actions are of necessity and not by choice. Monthly forecasts have little to offer beyond interest, because there is little a farmer can do at a moment in time to help matters which will come into prominence some weeks later. He is already committed to a course of action, and has no alternative but to wait and do his best when the time comes.

Seasonal forecasts are another matter, but here again, the weather of the next season is too close in time for most agricultural planning. Foreknowledge of summer conditions could affect choice of spring sown crop or variety. If the severity of the forthcoming winter is known, decisions can be made regarding autumn husbandry, but in general plans have to be laid about six months or more ahead.

Farming is always forced to take an operational risk regarding future weather. The immediate risk concerns the next crop, or the next farming year, but the long-term risk involves weather over the next 10—20 years. In other words, a form of climatic trend forecast would be of incomparable value to land-use planning especially in areas at the fringe of suitable agro-climatic zones.

The fringe areas can normally accept an adverse risk of 3 in 10, but when this increases to 4 or 5 in 10, then the probability of crop failures in two successive years becomes nearly a certainty and few enterprises can withstand such adverse circumstances.

The effects of weather

Before any use can be made of a weather forecast, the agricultural effects of the expected weather must be understood, so that work in agricultural meteorology which concerns itself with that precise problem is an important pre-requisite for the efficient use of forecast material. The forecaster should be made aware of the critical types of farming weather, both in general, and also whenever possible, in precise numerical terms. The farmer should have the effects of the weather explained to him, both as part of his general education, but also as part of his daily advice.

This means that it is not always advisable for weather forecasts to be communicated to farmers by meteorologists. It is often preferable for an agricultural adviser to interpret the operational meaning of expected weather in any public communication.

Pests and disease

The one type of forecast which a farmer or grower can do something about is that which concerns the incidence and intensity of attacks by pests and disease. Many such forecasts are principally based on past or present weather, but some do perforce involve a conditional sentence involving future weather. Phrases such as "especially if the present unsettled weather continues" or "unless the next few weeks are dry" add value to advice which may have to be given some days or weeks in advance of required action.

It is always advisable to include any form of weather prediction in such a provisional manner. There is always a degree of error in any macroscale relationship between a pathogen and the weather, and if to this is multiplied the degree of error inherent in a weather forecast, then the resulting degree of possible inaccuracy can become operationally unacceptable. No form of forecast is any use unless it is believed in, and if the credibility gap is too wide, not even a correct forecast will be acted upon.

Yields

The yield of a crop depends on the weather up to the end of harvest and beyond. It is therefore very difficult to make a yield forecast using past weather alone, even though past weather may at times have had an irreversible effect. The effect of the double uncertainty mentioned above is even more important in this context. With very few exceptions, yield forecasts have a relatively large radius of error, and at present there seems to be little hope of rapid improvement.

Publication and publicity

It is very difficult to devise a system of efficient communication between weather forecaster and farmer. There is a (very low) limit to the number of farmers who can consult a weather centre personally or by telephone. Press, radio and television all have their severe limitations. Time is limited, details have to be omitted even if they are available; the listener or viewer does not always understand or even hear the complete bulletin; press forecasts are inevitably read long after they have been issued.

Automatic recorded telephone answering systems seem to offer the best service on a wide scale, especially if the forecast can be functional, or in other words, phrased in terms of farming operations. This demands close co-operation between forecaster and agricultural adviser. The offering of a simple unsophisticated weather forecast, solely in meteorological terms, to the public has been likened to preparing a meal and then throwing it out of the window in the hope that some passer-by will recognize it as food and eat it.

Weather forecasts, even with the present limited accuracy and range, can be useful, but their best use depends on the degree of education of both meteorologist and farmer and the efficiency of intercommunication. Their helpfulness will always be limited by the helplessness and vulnerability of a farmer — there is often so little he can do.

SUGGESTIONS FOR FURTHER READING

Irrigation*

Smith, L.P., 1954. Calculating irrigation needs. *Sci. Hort.*, 11: 11—17.

-----

* The United Kingdom Ministry of Agriculture Bulletin No. 138, entitled "Irrigation", is being revised and is to be published late in 1974.

Smith, L.P. and Meads, D., 1961. Frequency of irrigation needs for potato crops. *NAAS Q. Rev.*, 54: 72—77.

Hogg, W.H., 1961. Irrigation needs for grass in England and Wales. *Univ. Coll. Wales, Aberystwyth, Mem.* 4: 28—31.

Hogg, W.H., 1967. *Atlas of Long-Term Irrigation Needs for England and Wales.* Minist. Agric. Fish. Food, London.

Hogg, W.H., 1969. Estimates of long-term irrigation needs. *Univ. Coll. Wales, Aberystwyth, Mem.*, 12: 11—19.

Shelter

Gloyne, R.W., 1954. Some effects of shelterbelts upon local and microclimate. *Forestry, Oxford*, 27: 85—95.

Gloyne, R.W., 1955. Some effects of shelterbelts and windbreaks. *Meteorol. Mag., Lond.*, 84: 272—281.

Hogg, W.H., 1956. Wind shelter for horticultural crops. *Agric., Lond.*, 62: 587—590.

Hogg, W.H., 1960. Measurements of relative shelter at Rosewarne Experimental Horticulture Station. *Expl. Hort.*, 3: 7—12.

Hogg, W.H., 1961. Measurements of the effect of shelterbelts at Stockbridge House. *Expl. Hort.*, 4: 62—67.

Hogg, W.H., 1962. Shelter screens at Luddington. *Expl. Hort.*, 7: 47—51.

Hogg, W.H., 1964. Lath shelter screens for lettuce at Stockbridge House, 1959—61. *Expl. Hort.*, 11: 23—28.

Hogg, W.H., 1964. The shelter needs of horticulture, methods of providing shelter and plant response. *Proc. Symp. Shelter Res., Aberystwyth, April 1962. M.A.A.F., London*, pp. 11—30.

Gloyne, R.W., 1964. Meteorological techniques in shelter research. *Proc. Symp. Shelter Res., Aberystwyth, April 1962. M.A.F.F., London*, pp. 45—66.

Gloyne, R.W., 1965. Some characteristics of the natural wind, and their modification by natural and artificial obstructions. *Sci. Hort.*, 17: 7—19.

Gloyne, R.W., 1965. Some meteorological aspects of shelter research. *Proc. 2nd Symp. Shelter Res., Edinburgh, Sept. 1964. M.A.F.F. London*, pp. 53—63.

Hogg, W.H., 1965. Studies in horticultural shelter. *Proc. 2nd Symp. Shelter Res., Edinburgh, Sept. 1964*, pp. 3—12.

Hogg, W.H., 1965. A shelterbelt study; relative shelter, effective winds and maximum efficiency. *Agric. Meteorol.*, 2: 307—315.

Hogg, W.H., 1965. Measurements of the shelter effect of land-forms and other topographical features, and of artificial windbreaks. *Sci. Hort.*, 17: 20—30.

McKay, W., 1969. Some effects of deciduous shelterbelts at Stockbridge House. *Expl. Hort.*, 19: 1—15.

Gloyne, R.W., 1970. Shelter and local climate differences. *Weather, Lond.*, 25: 439—444.

Gloyne, R.W., 1971. The assessment of exposure by the rate of disintegration of a standardised textile flag. *Rep. Joint Shelter Res. Comm. M.A.F.F., London*.

Housing

Smith, C.V., 1964. Animal housing meteorology, I. Ventilation and associated patterns of air flow; II. Rating of ventilation systems for animal houses; III. A quantitative relationship between environmental comfort and animal productivity. *Agric. Meteorol.*, 1: 30—41; 107—120; 249—270.

Smith, C.V., 1967. Airborne tracers in agricultural meteorology. *Meteorol. Mag., Lond.*, 96: 150—155.

Smith, C.V., 1972. Some environmental problems of livestock housing. *W.M.O. Tech. Note*, 122, 71 pp.

Storage

Smith, C.V., 1969. Meteorology and grain storage. *W.M.O. Tech. Note*, 101, 65 pp.

Glass cover

Smith, L.P., 1951. Temperatures under Dutch Lights. *Meteorol. Mag., Lond.*, 80: 50—52.
Gloyne, R.W., 1954. Temperatures under glass. *Sci. Hort.*, 11: 158—162.
Jackson, A.A., 1959. The eco-climatology of Dutch Lights, I. Meteorological assessment; II. The effect on growth. *Weather, Lond.*, 14: 117—123 and 155—162.
Hogg, W.H. and Ebsworth, H.L., 1960. Frost protection under cloches by use of irrigation water. *Expl. Hort.*, 3: 69—70.
Hogg, W.H., 1964. Frost prevention in Dutch Light frames. *Agric. Meteorol.*, 1: 121—129.
Smith, C.V., 1970. Calculated glasshouse light transmission, the effects of orientation of single glasshouses. *Expl. Hort.*, 22: 1—8.
Smith, C.V. and Kingham, H.G., 1971. A contribution to glasshouse design. *Agric. Meteorol.*, 8: 447—468.

Transport

Allen, W.M., Hebert, C.N. and Smith, L.P., 1974. Deaths during and after transportation of pigs in Great Britain. *Vet. Rec.*, 94: 212—214.

Forecasts

Smith, L.P., 1967. The world weather watch and meteorological service to agriculture. *W.M.O. World Weather Watch Planning Rep.*, 22: 28 pp.
Hogg, W.H., 1970. Long-range weather forecasts. *Agric. Lond.*, 77: 433—435.
Hogg, W.H., 1971. The weather forecast requirements of specific types of agriculture and horticulture. *Univ. Coll. Wales, Aberystwyth, Mem.*, 14: 99—114.
Hogg, W.H., 1971. Weather forecasting for agriculture. *Agric. Lond.*, 78: 352—355.
Gloyne, R.W., 1972. Long-range weather forecasts and the farmer. *Scott. Agric.*, 51: 261—267.

CHAPTER 7

The Modes of Agricultural Meteorology — Hazards

> "One cannot increase the size or quantity of
> anything without changing its quality."
> PAUL VALERY

> "Out of this nettle, danger, we pluck this
> flower, safety."
> WILLIAM SHAKESPEARE

FROST

The best way to combat frost is to avoid it. If frost-sensitive crops are planned, areas should be chosen which are least liable to frost incidence. Farm buildings should not be sited in areas of bad frost risk, unless a natural supply of cold air is required, as in the case of potato storage. The selection of frost-free sites, or areas with a minimum frost risk is a problem of agroclimatology, although it may often be necessary to take a series of minimum temperature observations within the area in question to supplement existing records.

Agro-meteorological help is needed when the site choice has been a bad one, or when even the best available site still has an unacceptable frost risk. It might be thought that frost forecasts would be of prime importance, but it cannot be emphasized too often that for a forecast to be of any practical use several difficult conditions must be satisfied. The forecast must be accurate, both in timing and intensity, for a precise small locality; it has to be made available to the person concerned in terms which he fully understands, and, most important of all, he must be able to do some something useful on receipt of such information. The frost forecast requirements of a farmer or grower are thus very demanding indeed, and they present immense difficulties to the professional forecaster.

Although generalized area warnings of frost are of limited use, they do indicate the degree of urgency and prompt the agriculturalist to make his own local decisions. Considerable advance warning can in fact be given regarding this degree of risk, because late spring frosts are much more liable to occur in May, if the preceeding months of March and April have been dry. This influence is clearly shown from the examination of past records.

For example, taking the mean of 29 stations in England, each with 30-year records, it was found that: if the rainfall in March and April exceeded 125 mm, then 32% of the following Mays had frost; if the two-month total was below 125 mm, the percentage rose to 57%; if it was below 100 mm, 59%; if below 75 mm, 64%; if below 50 mm, 75%.

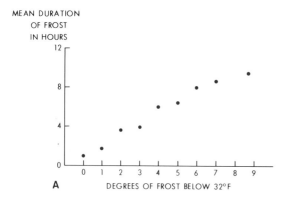

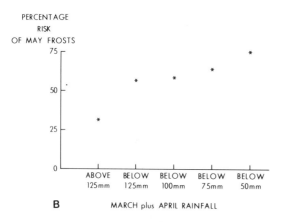

Fig. 21. A. Relationship between frost intensity and frost duration. B. March and April rainfalls and risk of May frosts.

Thus the general odds of the occurrence of damaging late frosts after a wet spring were about 2 to 1 against, but if the weather had been dry, then the odds shortened to 3 to 1 on (see Fig. 21B).

Night minimum temperatures vary appreciably over very small distances, and it is quite unreasonable to expect any forecast service to be able to give a correct estimate of frost for every small locality. There can be no alternative to experienced local judgement, but help can be given to those who wish to make their own forecasts and who are prepared to take a few simple on-site measurements of the weather.

One of the best ways is to prepare a guidance diagram which has built into it the local site characteristics, because it is prepared from past records on the same site. For example, if at 18h00 GMT the following observations are made for each evening in March, April and May: T = dry bulb temperature of a screened thermometer in °F; W = wind speed in miles per hour (which can be an approximate visual observation); C = cloud cover in eighths; R = number of days since last rainfall (with a maximum of 8); D = dewpoint in °F.

Using these data, a dot diagram is produced, with the factor $(T + W + C - R)$ on the y-axis, and D on the x-axis, and using a black dot for nights with a frost; a small dot in a circle for nights with only ground-, but not air-, frost; and a small open circle for nights following the observations which were free from all frosts. It will be found that the different type of entry on the diagram group themselves together, with the frosts nearest the origin. Curves can be drawn on the dot diagram separating in general the different type of night minima. This diagram can then be used in succeeding years to provide a reasonably reliable frost forecast peculiar to the site in question. All the local influences which an outside forecaster cannot know, let alone take into account, are incorporated on the evidence of the previous occurrences (see Fig. 22).

Clearly, any such system can be simplified or it could be refined still further by taking into account the wind direction or the behaviour of the barometer; all depends on the intelligence of the operator and the extent to which he needs the foreknowledge. Some growers with permanent antifrost installations rely solely on a thermometer with an alarm system which operates when, say, one or two degrees above freezing point is reached.

Agro-meteorologists have to concern themselves with the assessment of the efficiency of methods used to protect crops against frost, but it is not always appreciated that good husbandry techniques can help to lessen the frost risk. The only natural supply of heat at night is that which is conducted upwards from soil which is warmer than the air above. To improve this supply of conductive heat, it is desirable that the soil surrounding frost-susceptible plants should be moist, compact and free from weeds. There is a mass of supporting evidence for this ameliorative effect; in The Netherlands it has been shown that spring-ploughed land is more liable to frost than that which has been autumn-ploughed and subsequently settled; in Australia, a special allocation of irrigation water is supplied during the late spring frost risk period.

Not all non-meteorologists realize that night temperatures are almost always lower over grass than over bare soil, and that minima are lowest of all over a mulch which contains a high proportion of insulating air. A straw mulch in particular will provide conditions for a severe frost risk. Frost incidence will also vary with soil type, clay soils being the least liable and sandy soil being more prone to frost; a peaty soil, especially when the surface layers are dry, will produce the lowest night temperatures of all.

There are many methods advocated commercially for protecting crops against frost, many of which are reviewed in *W.M.O. Technical Note* No. 51. Most, if not all of them have limited efficiency and can be costly to operate;

on a really bad frost night the amount of the heat energy lost by the cooling earth can be greatly in excess of any palliative. When assessing their relative efficiency, great care must be taken in the field experiments to measure the temperatures accurately. In particular, the exposure of the measuring instruments must be strictly comparable. Different types of thermometer should on no account be used in the same experiment, except that standard screen measurements should always be taken to provide a dependable reference point.

An important factor in considering the economical use of protection measures is the duration of a frost. In this respect it is worth remembering that a

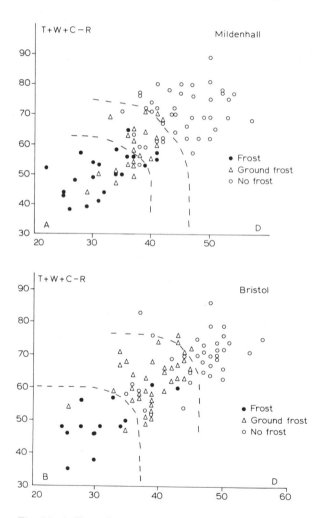

Fig. 22. A. Frost forecasting nomograph for Bristol (southwest England).
B. Frost forecasting nomograph for Mildenhall (East Anglia).

simple approximation to the number of hours of frost on any given night is the number of degrees F of frost below 32° plus one (see Fig. 21A). There is little doubt that the most efficient method of protecting crops against frost involves the use of continual fine sprays of water. Care must be taken that the spray system is such to ensure that the plant surface is always covered by a mixture of water and ice (with a temperature of 32°F), and yet the minimum water should be used, otherwise the ice load becomes too great. If the plant becomes covered with ice alone, then its temperature will drop rapidly below freezing and the damage will be severe. The plant, bush or tree must be strong enough to bear the load of ice without breaking and the soil should be of a type which will be capable of draining away rapidly the excess water. Experience has shown that up to 5°C of frost protection can be achieved by spraying methods, roughly twice as effective as most other methods.

Glass cover also provides a degree of frost protection (see p. 149), although some plastic covers offer very little. The combination of glass cover plus fine water sprays can be very effective indeed. If matting is laid over a glass cover it has to be removed in the morning, but an ice cover which keeps the outer surface at freezing point but not below, will be automatically removed by thawing in the sun of the following morning, thus saving many valuable man-hours of labour.

Gardening tradition recommends the use of sprays over frosted plants to induce a slow thaw the following morning and claim that this practice reduces damage. Little experimental evidence can be found to support this, and it is certain that if cells have been ruptured by frost no post-damage treatment will do any good. It may be possible, however, that a foliar spray in strong morning sunshine following frost will help to avoid a temporary transpiration problem, but the effect can only be marginal.

FOREST AND BUSH FIRES

Although weather plays a significant part in the occurrence and control of forest and bush fires, it is difficult for an agricultural meteorologist to carry out any research in the subject unless he is working in close collaboration with the forestry or fire-fighting authorities; he cannot work alone.

Three aspects of the problem are of particular importance. Firstly, the weather leading up to a state of high fire risk has a significant effect on the state of the trees or other vegetation, and also on the condition of the litter on the forest floor. Secondly, the weather may be the cause of the start of the fire, usually through a lightning strike, although cases of spontaneous combustion have been reported. Thirdly, the weather during the course of the fire, especially the wind force and direction and the incidence of rain, has a big bearing on the ease or difficulty with which the fire can be brought under control.

W.M.O. Technical Note No. 42 contains a review of the various aspects of the subject. It should be consulted before any work is undertaken. In countries where the fire hazard is a regular threat, fire indices which summarize the potential danger of fire are probably already in use; they may, or may not, contain meteorological parameters. There is little doubt that some of them could be improved by better use of meteorological knowledge and weather reporting services.

The forecasting sections of a meteorological service can also play a helpful role in the estimation of fire risks. Their regular and reliable reporting organization can provide data which could be fed into any system of fire rating assessment, making use of such factors as the number of days since the last rainfall, the minimum daily relative humidity or the day maximum temperature. Arrangements can also be made for the issue of specially designed forecasts of future weather, following the specifications laid down by the forest authorities. Such forecasts can be issued on a regular basis, or can form the basis of a warning system in which they are only issued when certain threshold conditions are exceeded.

During any fire-fighting operation, frequent telephone consultations between fire control centres and the nearest available forecast office can be of the greatest value, especially in regard to the possibilities of rainfall, increased wind strength, or large changes in wind direction.

On a minor scale, inconvienience to traffic and risk of accident on highways can arise if farmers are burning stubble in fields after the grain harvest. At least one such fire has been known to cause a multiple collision accident on a nearby motorway due to the dense smoke. Any system of making forecasts of wind speed and direction available to farmers under such circumstances cannot but have a useful effect. The same applies to any form of controlled burning of coarse grass or heather which is resorted to in order to promote new young growth.

Many fires are caused by incautious visitors to amenity areas in parks or nature reserves and can wipe out years of growth in trees or ruin an annual crop. Again the closest cooperation between the authorities and the weather experts can help to minimise the risk.

W.M.O. Technical Note No. 42, entitled "Forecasting for forest fire services", contains valuable information on these problems (see Appendix I).

METEOROLOGICAL HAZARDS

The term "hazards" includes all the types of extreme weather that can cause damage to crops, animals, and agricultural buildings. The role of the meteorologist is to identify the danger criteria. This being done, he can use climatological data to quantify the danger risk at any time and place. Protective action can rarely be taken against the major hazards, and the evasive action consists of either the selection of areas of minimum risk, or the selection of enterprises with minimum sensitivity to weather extremes.

Three categories of weather are liable to such extremes, namely, tempera-ture, precipitation and wind. These factors cannot always be treated separa-tely, because it is often the combination of two adverse factors which creates the worst type of hazard. In addition to the direct effects of extreme weath-er, there is also a need to consider the secondary effects.

Many of the more common hazards form the subject of other chapters. These include frost and high temperatures, drought and drainage, wind and shelter. The major hazard which arises because of excess precipitation is the incidence of floods. Flooding is, strictly speaking, a hydrometeorological problem, but it involves many considerations common to agro-meteorology.

Broadly speaking, there are three main types of flood. The first type is due to intense local rainfall, wherein a very large amount of water is precipi-tated over a relatively small area within a short space of time, such as in a heavy thunderstorm. Under these circumstances, agro-meteorological factors play a relatively minor part, except that the soil moisture status and the type of ground cover may have some modifying effects. The flooding is caused by the fact that the rate of arrival of water at the surface of the earth is much greater than the infiltration rate into the soil. This infiltration rate, the speed at which the water can penetrate into and through the soil is higher when the soil is below field capacity (especially if the surface layer is moist), and least when the soil is saturated. Initial infiltration is also slow when the surface is very dry, under which circumstances there is liable to be a rapid surface run-off. With short-duration high-intensity rainfall, this reluctance of the dry soil to begin the process of wetting may be very important. In regard to the type of ground cover, forests of trees with extensive root systems can facili-tate the absorption and downward movement of water in the soil; similarly a cover crop of grass is more helpful than row crops with incomplete ground cover. Roads and hard standings in and around the farm are, of course, the least helpful ground cover of all. A downhill road leading to a farm can become the equivalent of a raging torrent if heavy rain is falling on the hills, especially if the roadside ditches are choked.

The second type of flood occurs when snow begins to thaw rapidly. This can often be the case when there is a rapid rise of temperature accompanied by rain. Warm rain will melt snow more quickly than hot sunshine, most of which is reflected by the white surface. Such floods will have severe agricul-tural effects, but there are few agricultural influences on their intensity, except that snow in forests is likely to thaw less quickly and again the large tree root systems offer better chances of infiltration.

The third type can be called the autumn or winter flood which is caused by prolonged periods of rain. Although the rate of rainfall, as compared with the intense rates of the heavy storm, may be quite low, the total precipita-tion over a period of one or more days may be considerable. If there is a deficit below soil moisture capacity, it is often possible for such rain to be absorbed safely into the ground, and apart from local waterlogging in poor drainage areas, the flood danger will be low. On pronounced slopes, some surface run-off will be inevitable, but will rarely reach dangerous proportions. The situation is very different when soil capacity has been reached, and almost all serious autumn and winter flooding occurs after the date of return to capacity. Agro-meteorological estimates of the soil moisture status, and especially the calculation of this return date, are therefore of the greatest importance in the assessment of flood risks (see p. 47).

In many countries, it is the river authorities and not the metoeorological service who are responsible for flood forecasting. Nevertheless, information regarding soil-moisture deficits and return dates can be provided by agrometeorologists or hydro-meteorologists, and this information is valuable to farmers and the responsible river authorities.

Apart from any question of flooding, heavy rain can damage crops. The impact of a fast moving drop of water on a delicate crop can ruin it commercially and seriously affect its growth and subsequent yield. Hail-storms can cause even worse damage, especially to crop-cover structures of glass or plastic. Little can be done to prevent such damage, as hail prevention is still an imperfect operation, and claims to be able to disperse clouds producing hail are more an article of faith than a proven scientific fact. Forecasts are of no help at all, because even if they were accurate in space and time, next to nothing can be done to protect the crop or the glass.

Buildings can be protected against lightning by the provision of lightning conductors, although few farm buildings, with the exception of grain silos, are tall enough to warrant such precautions. The chief lightning damage suffered in agriculture is the loss of livestock which may be sheltering, clustered together, under a large tree. If the tree is struck, most of the animals around its trunk and under its canopy are liable to be killed. A wire fence is also a hazard to stock in such conditions, against which there is no defence.

Snow is a special form of precipitation which by its very weight can cause damage to shrubs and trees, but again there is very little to be done about it. Roads can be kept free from serious snowdrifts by the sensible siting of snowfences, and on upland farms in particular such regular winter precautions are worth their while to keep open the necessary communications. Upland stock, especially sheep, can be buried in snowdrifts, and will die unless found and released. Accurate forecasting of the incidence of snow and especially the liability to drifts is of the utmost importance to give the shepherds warning of the possible dangers. Regrettably, snow is not the easiest of features to forecast and only a small change in conditions can turn harmless rain into lethal snow.

A climate which involves a high frequency of moderate to strong winds can be improved by the careful planning of wind-breaks or shelter-belts, but the major wind hazards are encountered on the occasion of gales. The first gale of late summer or early autumn is often (incorrectly) thought to be the most severe, but this is probably because it is liable to cause the most damage. Not only are crops more susceptible at this time, and indeed an untimely gale could bring down an entire apple crop, but also the first strong winds of the winter half-year will inevitably find out all the weak spots in farm buildings and the half-dead branches and trunks of trees. Again there is little that can be done at the time, but good maintainance can forestall a lot of damage and old trees should be felled before they reach the dangerous state.

Combination hazards

A coincidence of two weather hazards in time and place can have very serious results. For example, coastal flooding can occur if a high tide at springs coincides with an on-shore gale. A triple coincidence can be even more serious; a river running bank-high with the run-off from recent rains can meet a gale blowing into the river mouth plus a high spring tide, and under such conditions the normal flood precautions can be totally inadequate. The authorities or forecasting services cannot be blamed unduly, because such a combination of circumstances may only occur once a century or less, and it is hardly economic to plan expensive defence works against such a contingency. It is far cheaper to pay for the damage. The prime requisite is to have a warning system which will prevent the loss of life.

Other damaging combinations of events occur more frequently. Wind and rain, or rain and bitter cold are not uncommon; both cumulative effects can cause loss of production or even loss of life in outwintered stock, especially pregnant ewes. The coldest nights are often those with least wind, which is just as well because the combination of a biting wind and low temperatures can cause real problems even for livestock which are housed, and will cause a severe stress to any in the open.

Snow by itself is a problem, but snow with gale force winds will cause severe drifting out of all proportion to the total amount of precipitation. Not least of the blizzard hazards is the fact that it will reduce visibility even in daytime, to almost zero. In such "white-out" conditions it is extremely difficult to find any stranded livestock and to give them food or other aid.

Secondary effects

In addition to direct damage, the secondary effects of weather hazards are complex and numerous. The first obvious example is that of erosion, which is a subject in itself. The extreme cases of slip erosion, such as landslides, or the wholesale movement downslope of old dumps of industrial or colliery waste cannot be termed part of agro-meteorology, and the same may be said of avalanches. Extreme cases of wind erosion lead to sandstorms which are mainly experienced in desert areas, but rising sand can damage crops on the fringe of arid areas and soil "blowing" can be serious in many arable areas on light land.

On-shore gales do more than cause structural damage or be a contributary cause of river flooding. They transport minute particles of salt from the sea spray for many miles inland and this, like sand, can bring about serious damage to trees and crops. Glasshouses near the sea may have to specially cleaned of salt deposits after a gale.

A less obvious, but nevertheless important secondary effect of gale damage is that bruised plant and tree tissue can provide entry points for subsequent infection by a variety of pests and diseases. Fire-blight, for example, has been shown to spread more effectively following occasions of strong winds (or hail) and heavy rain.

The necking of a cereal crop, that is to say the half-breaking of the stem near the ear is probably another effect of hail or heavy rain. The plant may survive but the microclimate for ripening is significantly altered. The lodging of a cereal crop, the beating down of the crop stand to ground level, is due to a combination of rain increasing the weight of the ears plus a strong wind to distort the crop structure and the effect of the physical impact of the raindrops acting like a battering ram. If the subsequent weather is dry, the crop can at times be harvested successfully with little loss in yield; there is even some evidence that ripening is accelerated. If wet weather follows lodging, sprouting can occur in the ear and the crop become a total loss. Strong winds close to harvest can cause shedding of the grain, but this is more a direct than a secondary effect. Strong winds and high temperatures can provide the ideal conditions for the spread of a disease such as mildew, the spores being released by shaking.

A grass pasture which is subject to excess rainfall leading to temporary waterlogging of the soil could subsequently recover and be little the worse, but if cattle are kept on such fields, then poaching will result. Their feet will tread the grass into the mud or wet ground, consolidating the soil into a state without adequate air or oxygen and preventing recovery for a long period of time. Wheeled traffic over soil containing too much water can also damage the soil structure, the effects of which can be long lasting and may need special treatments before recovery.

Although little can often be done to protect crops against major meteorological hazards, methods of protection have been developed for many less extreme conditions, and *W.M.O. Technical Note* No. 118 includes an extensive bibliography concerning such methods (see Appendix I).

SUGGESTIONS FOR FURTHER READING

Frost

Lawrence, E.N., 1952. Frost investigations. *Meteorol. Mag., Lond.*, 81: 65—74.
Lawrence, E.N., 1952. Estimation of weekly frost risk using weekly minimum temperatures. *Meteorol. Mag., Lond.*, 81: 137—141.

Smith, L.P., 1953. Estimating the frost risk of an orchard site. *NAAS Q. Rev.*, 19: 291—295.

Lawrence, E.N., 1956. Minimum temperatures and topography in a Herefordshire valley. *Meteorol. Mag., Lond.*, 85: 79—83.

Hogg, W.H. and Ebsworth, H.R., 1960. Frost protection under cloches by the use of irrigation water. *Expl. Hort.*, 3: 69—70.

Smith, L.P., 1964. A simple method of frost forecasting for use by growers. *Proc. Int. Hort. Congr., 16th, Brussels, 1962*, pp. 175—179.

Hogg, W.H., 1964. Frost prevention in Dutch Light frames. *Agric. Meteorol.*, 1: 121—129.

Hurst, G.W., 1966. Temperatures in the forest of Thetford Chase. *Meteorol. Mag., Lond.*, 95: 273—279.

Hogg, W.H., 1967. Frost types and incidence; some horticultural and agricultural implications. *Univ. Coll. Wales, Aberystwyth, Mem.*, 10: 57—59.

Hurst, G.W., 1967. Frost variations in Thetford Chase. *Univ. Coll. Wales, Aberystwyth, Mem.*, 10: 143—157.

Hurst, G.W., 1967. Further studies of minimum temperature in the forest of Thetford Chase. *Meteorol. Mag., Lond.*, 96: 135—142.

Hogg, W.H., 1968. Analysis of data with particular reference to frost surveys. *Proc. W.M.O. Reg. Training Sem., Wageningen, 1968*, pp. 343—350.

Hurst, G.W., 1969. Studies of temperature in the forest of Thetford Chase, spring 1967. *Meteorol. Mag., Lond.*, 98: 39—49.

Cochrane, J., 1969. A relationship between minimum air temperature and duration of frost in late spring. *Meteorol. Mag., Lond.*, 98: 138—140.

Hogg, W.H., 1971. Spring frosts. *Agric., Lond.*, 78: 28—31.

Hazards

Spence, M.T., 1955. Wind erosion in the Fens. *Meteorol. Mag., Lond.*, 84: 304—307.

Spence, M.T., 1956. Damage to crops by lightning. *Meteorol. Mag., Lond.*, 85: 387—389.

Roswell, H., 1956. Damaging hailstorms. *Meteorol. Mag., Lond.*, 85: 344—346.

Smith, L.P., 1964. Weather hazards in agriculture. *Univ. Coll. Wales, Aberystwyth, Mem.*, 7: 1—9.

Gilbert, E.G. and Walker, J.M., 1966. Tornado at the Royal Horticultural Society's garden, Wisley. *Weather, Lond.*, 21 (6): 211—214.

Hurst, G.W. and Rumney, R.P., 1971. Protection of crops against adverse weather. *W.M.O. Tech. Note*, 118, 64 pp.

CHAPTER 8

The Market-Place of Agricultural Meteorology

"Man was not born to solve the problems of the universe,
but to put his finger on the problem and then to keep
within the limits of the comprehensible."
J.W. VON GOETHE

"No man is exempt from saying silly things, the
mischief is to say them deliberately."
M.E. DE MONTAIGNE

"To write simply is as difficult as to be good."
W. SOMERSET MAUGHAM

VERBAL PRESENTATION

The opportunities for the spoken presentation of the results of research or investigational work may be few, but even if they are relatively numerous, this is no excuse for not making the most of every occasion. Some scientists are very bad at explaining their work verbally or may be shy at speaking in public, and certainly apart from the few gifted natural speakers, it is an art which often has to be learnt the hard way, profiting from each mistake and improving with each occasion. The following suggestions may be found useful and help to prevent good material being spoilt by indifferent presentation.

Consider first a meeting where all present share the same native language.

The audience

The first important point is to consider carefully the nature of the audience, which will materially influence the contents of the speech and the style of delivery. In some respects it might be thought that it would be easiest to speak to a meeting of scientists of the same discipline and working on roughly the same subject, but it must be remembered that they are likely to be the most critical, not necessarily unkindly, but because they are the most knowledgeable. In such circumstances much of the elementary details can be omitted, because they are already generally known, and nothing is worse than providing a series of glimpses of the obvious. Even so the finer details of a new line of argument cannot be smudged over, for the audience will be quick to perceive an over-optimistic conclusion or a sequence of doubtful logic. The speaker can start from what he knows his audience already understands and build confidently on what normally should be a solid scientific basis. The only exception arises when the paper attempts to confute an accepted theory or hypothesis. Under these circumstances it is best to present the new evidence first, then show how this does not fit with existing concepts and finally present a new type of explanation or procedure.

Modifications must be introduced if the audience consists of scientists of the same discipline, but not wholly working in the same type of subject. Here it is impossible to assume that everyone is aware of the knowledge and events leading up to the present paper. The start must be made a little further back in scientific time and concepts familiar to the co-worker cannot be dismissed in a casual phrase or "in-word". For example, the phrase "potential transpiration" should not need an explanation in an audience of agro-meteorologists, but it does demand a sentence or two of definition to listeners who may be meteorologists but who have never worked in hydrological or biological problems.

This modification process must be carried still further when speaking to scientists of other disciplines. It must firmly be kept in mind that what is elementary to a meteorologist may be completely unknown to a biologist, and vice versa. Words which are commonplace to a "home" audience, cannot be used in an "away" match without carefully considering whether there is a possibility of misunderstanding. While it is perfectly fair to take the standpoint that "a word means exactly what I intend it to mean, neither more nor less" it is essential to explain the intended meaning. Care must also be taken not to make any major blunders in regard to the associated sciences. An attempt to speak in terms of the other science is excellent, but it must be a successful attempt, otherwise there is the risk that easily recognized errors in a subject the listener is familiar with, will unjustifiably imply a lack of quality of the speaker in his own science. Wherever possible, the subject should be tackled from the point of view of the listeners and the science of the speaker should be fitted into this framework. In some cases, however, the sequence can be reversed; the main purport of the paper can confine itself to one science, and the implications for the associated subject can be brought in at the end.

A completely different set of problems arises if the audience is composed of non-scientists. Here the point of departure must be the interests and mental capacities of the audience. The simplification of a scientific subject is a very difficult process, and it may be necessary to sacrifice strict accuracy for the sake of clarity. The language used must be that of the listener and although the purpose may be one of scientific education in its broadest sense, it is essential to retain interest and to avoid the danger of confusion by attempting too intricate a talk.

Title and summary

Considerable thought should be given to choice of title; it should be chosen so as to attract the correct type of audience and yet not be mislead-ing. Possibly the best solution is to select a title and sub-title, wherein the title is brief and the sub-title more explanatory, in which case the title should be six words or less and the sub-title consist of about a dozen words. According to circumstances either or both can then be used.

The summary needs very careful thought. Like an after-dinner speech (or a miniskirt) it should be long enough to cover the topic but short enough to be interesting. If the title and/or sub-title has correctly stated the topic, a summary can conveniently be compressed into three sentences. The first sentence states the date or type of experiment: the second describes the methods of analysis; the third provides the outstanding results. Such a sum-mary is intended to be circulated to the potential audience, and therefore a vague or non-committal statement is almost insulting, and on the other hand an overlong and detail-crammed dissertation is equally likely to be ill-received. It has much in common with the abstract of a written paper but it should contain much less numerical details.

Duration

It is important to know well in advance how much time is available for delivery of the paper (or the length of a popular lecture), and it is even more important to keep to this schedule. At a large meeting in particular, it is ill-mannered, to say the least, to over-run the allotted time; moreover it is a sign of incompetence. It is perfectly easy, with a little trouble and a wrist-watch, to time the length of a spoken communication. Such a rehearsal should be carried out anyway, preferably more than once. It may even be true to say that the more extempore or more informal the speech, then the more time is needed for rehearsal. The fact that a previous speaker has exceeded his time allowance, does not always mean that his successor is forced to shorten his contribution, but in some circumstances he will earn grateful thanks by doing so.

At some meetings the time allotted includes an unspecified time for dis-cussion. It is often wise to make the utmost use of such a period; certainly it is discourteous to leave no time for discussion at all. Without in any way commending the wholesale "planting" of questions in the audience, a little priming by arrangement is often useful. A good chairman can often start a useful discussion with a carefully worded query or by calling on a speaker from the floor.

Contents and style

It must be realized that even with a good speaker and a good audience there is a speed limit for mental digestion. It is far better to put over one point well than to try and convey four or five new ideas and only leave the listeners in a confused state of mind. In other words, "ear-bashing" is unacceptable. As a general rule it is wise to restrict the contents to about one major point per five minutes in a short paper, and a maximum of 8 or 10 in an hour; more than that will be wasted.

A good opening sentence and a concise final statement are both most helpful, even if it is cynically thought that they would be all that the poor listener would remember.

The style of presentation is essentially a personal affair, but it must be remembered that spoken style is vastly different from written style even in the same individual, or at least it should be. A good rule is therefore: do not write a paper and then read it aloud exactly as written. The only exception is when speaking in a language foreign to you, when you lack the necessary fluency to speak in an extempore manner. Extempore speaking is an art which takes a long time to acquire, and it also demands a great deal of preparation. Reading a written paper is much easier for the speaker but much less satisfactory for the listener. If you lack sufficient confidence and experience to speak without repeating each written word, then make sure that the script is one which has been typed from dictation or from a dictaphone, so that a speaking style is retained.

One reason for this is that when the recipient is reading a communication via a printed page he can do so at his own pace; he has time to re-read a sentence or to refer back to a previous paragraph. The recipient who is listening to a talk or lecture has no such advantages; he is forced to try and understand the information at the pace set by the speaker. The spoken pace must therefore be slower and the steps of logic carefully thought out and clearly expressed. The precise pace must vary with the audience, and the experienced speaker can sense from the attitude of his listeners and the atmosphere in the room whether he is going too fast or too slow, not just in terms of words per minute, but rather in terms of logical progress. Most important of all, he should be able to detect a note of bewilderment or confusion, so that he appreciates the immediate need to expand in detail or to reiterate the logic. The speaker should also be sensitive to incipient boredom, and once he feels he is losing the attention of his audience, he should cut his losses and switch forward to his next important point.

The extent to which humour can be introduced into a spoken presentation varies with the speaker and the audience. If it is decided to attempt to enliven the proceedings with some form of humourous reference it must never be one which has long outlived its use. Clichés are bad at all times, but a joke which has been around long enough to be classified as a cliché can only introduce a sinking feeling, especially in a scientifically intelligent audience. Popular audiences, on the other hand, sometimes seem to enjoy old jokes, and positively feel cheated if they do not appear, subsequently filling the gap by introducing them into their questions at the end.

On the other hand, a sudden departure from a formal urbane style into a more forceful rhythm, including a short aphorism or appropriate metaphor, can often have a very useful effect. A few memorable words or even a paradox, will stay longer in the memory than a more classical sentence. It is a technique which has to be used sparingly, because the effect decreases with repeated use, nevertheless it can make the reputation of the speaker.

Visual aids

A blackboard and chalk is now being supplanted by writing on perspex and some form of light projection, which is just as well as the chalk dust will almost certainly cause the speaker to cough at some stage in the proceedings. Both types of visual aid are still valuable when speaking to small audiences or when answering questions at a larger meeting. In general, however, 35-mm slides are the most common illustrations used by speakers.

Such slides must be very carefully designed. They should *never* contain column after column of small figures. The audience on the front row may be able to see the figures but never have time to digest them; on the back row the figures cannot be seen at all. The essence of success with a slide is simplicity and if the diagrams become at all complicated it is very advisable to use two or more colours. Any words on the slides should be in large block capitals and should always be written horizontally. The attempts of an audience to read a caption written vertically along the *y*-axis may not lead to a broken neck, but it certainly engenders a quite unnecessary irritation.

Try to avoid a disjointed use of slides; a lecture in which the house lights are continually going on and off, or one in which the speaker uses three slides and then returns to the first one, then back to the fourth and so on, is very difficult to concentrate on. There are really two alternatives; either speak to a succession of slides throughout, or use the slides at the end as a form of recapitulation. The whole aim is to make it easier for the audience, and constant disruptions are the reverse of helpful.

Vocal aids

In a small room or with 'a small audience, a microphone should not be necessary. In a larger auditorium, it is inevitable, so the speaker must come to terms with the system. If the microphone is hung round the speaker's neck, there should be no problem, except to remember that there is no longer any need to pitch your voice to the back of the hall. If, on the other hand, it is fixed to the rostrum, care must be taken to speak evenly with no great variation in volume, and to remain at approximately the same distance from the instrument. This can be incapacitating to some types of speaker who like to move around the platform and can be very difficult to use when slides are being shown. The speaker turns to look at the illustration away from the microphone and his voice may be lost. A good tip is to move away from the screen and not towards it, thus ensuring that the voice carries sideways across the microphone. If it is inevitable that the microphone is somewhere behind you then your voice must be raised to compensate the drop in input. Most good speakers hate microphones except when addressing very large audiences, which leads us to the type of meeting where more than one language is necessary.

International meetings

When no interpretation facilities are available a considerable adjustment in style is necessary when addressing an audience wherein some are listening to a language which is foreign to them. To begin with speech must be slower, and care must be taken to select words and constructions so as to avoid misunderstanding; for example, an English double negative can be very confusing. If is often wise to repeat, in paraphrase, any important phrase or sentence. If the meaning is imperfectly understood on the first occasion, the second set of words should put things right. Remember that there are small, but sometimes very important, differences between English and American; remember also that words in English and French which look the same are almost certain to have slightly different meanings.

If simultaneous interpretation facilities are available, a microphone will always have to be used and used correctly. Professional interpreters reach a very high standard of competence but they are often unused to the technical terms of a scientific discussion. At formal meetings, when the speaker is using a script, they must be provided with copies well in advance. Where no script is used, it is always wise and helpful to check whether the interpreters are fully at home with certain technical terms. The best interpreters rise to the challenge of an unscripted speech, but the speaker must realize the strain under which they are working, must speak slowly with longer pauses between sentences. The German—English interpreter who appeared to have dried up completely was perfectly justified when he explained that he was "waiting for the verb" which would come at the end of the German sentence and near the beginning of the English version.

It is always wise to watch the interpreters in their soundproof boxes, if this is at all possible. If they are calm and relaxed, the speaker is probably going well; if they appear to be speaking rapidly and show a degree of agitation, it is time to slow down and give them a chance. A brilliant interpreter will even relish your jokes, if you give them a degree of warning, using not merely the strict literal translation, but the equivalent jest in the other language. The best laid plans for avoiding interpretation misunderstandings can, however, go astray. At an Anglo—Spanish meeting, for example, a full explanation to the interpreter of the meaning of the word "cross" in cattle breeding, resulted in the confident use of the Spanish equivalent of "Zebu—Holstein bastards".

It is worthwhile taking the utmost trouble regarding interpretation and translation at international meetings, especially at formal inter-governmental occasions. Some of the longest and most bitter arguments have at times arisen because of a slight misunderstanding of a key word. At symposia and conferences where there are a great number of international contributors, it is difficult to keep track of the standard of interpretation. Perhaps the best solution is a friend in the audience who is willing to listen in to the paper in the alternative language and who can then put any mistakes right during discussion time.

One other pitfall should be avoided, and that is the unconscious use of a current euphemism or "pillow-talk" of another language; only one consolation can be offered to the innocent who falls into such a trap, and that is he will pass into the immortal (and immoral) repertoire of translation jokes.

At a small meeting, consecutive interpretations can be an advantage despite the fact that it takes more time. The speaker must use short sentences and long pauses, but he does have the immediate chance to verify the translation if he has any knowledge of the second language. In any case the recipients can check him immediately if they do not understand.

International communication through interpretation can be very mentally tiring for all concerned, which is yet another reason why the speaker must be aware of the difficulties and the hazards and do everything possible to communicate clearly.

Broadcasting and television

There are some people who will be prepared to speak at a meeting, but who lack experience and confidence when dealing with the microphone, the tape recorder or the television camera, quite apart from those who hate public speaking anyway. Perhaps they miss the moral support of the audience or perhaps they are only too well aware of the apparent ability of both media to expose the inadequate or the insincere. Only two types appear safe, the man of character who knows his subject and the superb actor or actress who has learnt his (or her) part.

Broadcasting (on radio) falls into two general categories, the interview and the talk, although a general studio discussion falls somewhere between the two. With a little care and skilled planning in regard to both content and timing, it should be perfectly possible to construct a convincing interview. The timing is important to avoid the necessity for cutting and possible distortion of sense. Although the questions may be arranged, they should not be obviously synthetic, so that in answering the first question, the speaker should end with a sensible cue for the second question and so on. A good interviewer will welcome such co-operation and moreover in the somewhat claustrophobic studio can do much to help the beginner over any nervousness. Advice about rehearsals is difficult, because often the rehearsal, unless it is disastrous, can be better "radio" than the subsequent "take". A stumble over words is not a disaster, but evidence of muddled thought and semi-incoherence ceases to be human error and becomes downright incompetence.

A broadcast talk is very much a solo effort; as in public speaking, an extempore contribution based on hours of careful thought, is often far more convincing than a carefully read script. The great art is to forget the microphone is there at all and speak to a real or imaginary individual sitting opposite, and yet at the same time remember that the voice level should not vary unduly. Like many other things, ability to broadcast improves with experience, but even the beginner need have little to fear if he really believes in the truth and importance of what he is saying.

Television studios, to the uninitiated, are probably more upsetting than the radio equivalent. A radio studio, after all, is recognizable as a room with one glass wall behind which sit the engineers and their equipment and it is possible to forget their function. In television the background is only too obviously a small film set, and the speaker faces a battery of cameras and other disconcerting material which only serve to emphasize the artificiality of the occasion. Furthermore, the camera can be ruthless in exposing the deficiencies of the speaker, and there is generally little time to settle down and gain confidence. With good visual aids the situation is eased because much of the time the speaker is off camera and can relax. To have a simple but important story, to try to forget the surroundings, to pretend to speak to an individual and not orate to millions are probably the keys to a good performance, but they are easier to say than to do.

WRITTEN PRESENTATION

The truth is often best seen through the eyes of a poet. Robert Graves, for example, in his book "The Crane Bag" tells us: "Honest writers try to give their readers as little mental fatigue as possible. The most frequent cause of badly written English is a confusion of mind due, if not to ignorance, either to emotional stress or dishonesty." He also puts in a timely protest against those who "load their writings with the longest and least necessary words, in or out of the dictionary, which produce a mental fatigue encouraging readers to suppose that the theme and its handling are too awe-inspiring to dispute."

Clarity, in other words, must come before confusion, even before impressive confusion. All the disadvantages of the transient nature of the half-heard spoken word have been removed. The reader is anxious to learn, he is willing to read and reread the communication, so there is no excuse whatsoever for making his task more difficult.

The extent to which a scientist is capable of good writing depends partly on his own character and partly on the type of educational system he has experienced and survived in his youth; it also depends on the extent of his own reading, both of scientific and non-scientific books. The ability to write well rarely occurs quite as spontaneously as the ability to speak well, and it may take many years of practice to acquire a fluency of style. Moreover, writing styles have to be adjusted to suit the type of publication, in just the same way as styles of speaking must be appropriate to the type of audience.

Choice of publication

Scientific journals vary considerably, not only in standard but also in quantity and quality of readership; a prestige publication may, for example, have only a relatively small circulation, but this is not invariably true. Once a paper has been printed in any but the most insignificant of publications, then it is fairly certain that the title and summary will be included in the relevant series of scientific abstracts. This provides a chance of coming to the attention of a much wider scientific public, but even so it is advisable to try and select the most suitable journal, if only for the fact that it gives you more chance of being accepted.

Journals tend to sort themselves out under disciplinary headings; the number of publications which cater for interdisciplinary work are few, although in this respect agricultural meteorology is perhaps more fortunate than some other joint subjects. Papers written for scientists of one discipline and dealing with matters concerning another science have to be carefully written. For example, a contribution of weather and plant disease would need a different treatment when written for meteorologists than when written for pathologists.

In general, the writer should try to select the most appropriate channel of publication and should aim at the best quality journal that his paper deserves. He has, however, to be realistic; it is no use entering a light-weight in a heavy-weight contest, and alternatively it is not sensible to publish a very important contribution in a little known journal.

The preferred journal should be chosen, with one or two alternatives, before the paper is written. This being done, the journal itself should be studied to learn its style of presentation, normal length of contribution, and the instructions to contributors (often available at the end of an issue) carefully read and obeyed. There is absolutely no point in irritating an editor at any stage. He will normally be your friend, but he has a very difficult and time-consuming job to do. A paper for publication is therefore a "custombuilt" or "tailor-made" affair, it should not be written first and then hawked around to find a suitable customer.

Contents

Style of writing may be an intensely personal affair, but the sequence of contents in a paper is determined by the subject and designed for the convenience of the reader. It must be a logical sequence; a haphazard meander through a subject is all too often evidence of a butterfly mind, and will cast doubts on the validity of the contents.

The international journal *Agricultural Meteorology* (Elsevier, Amsterdam), for example, recommends in their "Guide to Authors" the following sequence in manuscripts, or more strictly speaking, typescripts: (a) Title; (b) Name of author and his (her) affiliation; (c) Abstract (or summary); (d) Introduction; (e) Methods, techniques, material studied, and area descriptions; (f) Results; (g) Conclusions; (h) Acknowledgments; (i) References; (j) Tables; (k) Figure captions.

The categories (a) and (b) are self-explanatory and need no amplification, but the remainder can now be examined in more detail.

Abstract

This is almost the most important part of the paper, because it is the only part seen by some in a collection of such abstracts, and in any case it will be the first thing a reader turns to, so that he can decide whether it is worth his while reading further. Remember the cynical phrase "Everyone writes papers, but nobody has time to read them."

The second half or even the major part of an abstract must provide a concise summary of the results; the first part should contain details of the material used or data, and refer to the area of consideration. This is important because an excellent result which has been obtained in one type of climate or in one set of circumstances is not necessarily suitable for universal application; like some delightful local wines, it does not "travel well". Numerical values can be included, and indeed in some cases, they are essential. Provided that adequate details are included a short abstract is preferable to a long one; it may even be thought that the importance of a paper varies inversely with the length of the summary. In any case, one trenchant sentence often carries more weight than a long paragraph. The inclusion of translations of the abstract into other languages is a most helpful practice, and should be adopted if at all possible.

Introduction

A good introduction should contain two things; firstly a brief review of previous work in the subject and, secondly, some explanation of why the present work was undertaken or why it is important and/or potentially useful. Again, length is to be avoided; brevity is not in this case the soul of wit, but it certainly is conducive to appreciation. The experienced and knowledgeable reader will know about the previous work anyway, but it is important to include some short reference to it, so as to show that the writer himself is not working in a mental vacuum or wearing metaphorical blinkers.

Data and methods

Sub-paragraph (e) in the above-mentioned list could perhaps have been written in the reverse order. The area studied, the data used or the measurements made should be specified first; methods and techniques included at this stage should only refer to the system of measurement or data extraction. The methods of analysing the data can be outlined in subsequent paragraphs, although it is possible to include information of this type in the presentation of the results.

Measurement techniques need very careful explanation, including the use of suitable diagrams or photographs. This is important because other workers may wish to repeat the experiments elsewhere, and also the conditions of measurement have to be fully understood before it is possible to assess the meaning of the results.

The data used must also be correctly described. This is a simple matter when standard meteorological data are being quoted, but biological data is rarely as internationally standardized and the reader must be left with no illusions concerning the representativeness of the material.

Results

Here is contained the real substance of the paper and clarity is essential. The results should be presented step by step in a series of almost independent sections. This sequence may, in fact, be the way in which the original work was done, or else, by use of hindsight, the way in which it should have been done. In any case the sequence must be logical, each successive section building as far as possible on that which has gone before. This may not always be possible as the problem might have been attacked from different angles, but the sequence of thought must be well designed to lead the reader through the paper without too much change of pace and direction.

If formulae are being established or hypotheses advanced, it is highly desirable to include details of independent checks. For example, given 10 years' data, it is wise to deduce a relationship on part of the range and test it on the remainder, even if this is only one year.

If several tests are available, it is totally dishonest to quote only those which give answers supporting your case. It is not only the hiding of the whole truth which is unforgivable but also the relative certainty of being found out which can lead to permanent damage to a scientific reputation. A suspicion of selective evidence can prejudice the success of a whole new concept, sometimes completely unfairly, and it is far better to admit an occasional failure than to leave it to others to discover the limitations and then condemn the whole.

Conclusions

In some papers this section comes under the heading of "Discussion" which is a more explanatory title, even though some conclusions are stated at the end of the discussion. Here again, honesty is essential and remember "Science is a first-rate piece of furniture for a man's upper-chamber, if he has commonsense on the ground floor". There is no sense, common or otherwise, in making unqualified extravagant claims during a written discussion, although similar statements in debate may serve as a useful mental jolt. The limitations inherent in the results must be recognized and stated; the objections to the inferences should be anticipated and, as far as possible, dealt with. This does not mean that obvious fallacies should be quoted simply for the purpose of rejecting them.

It might be said that a first-class paper has silenced any discussion and demonstrated its own conclusions to the mind of the reader before this section of the paper has been reached, but there are few contributions which reach this standard of excellence. The important aim is to convince the reader that the author has something valuable to say, but that he is under no illusion of omniscience and is aware of its possible weaknesses.

To a certain extent the style of presenting conclusions is a function of the country concerned. In Britain, there has for many years been a tradition of understatement, which therefore carries more conviction than a more immodest claim. In some cases this has been carried to extremes and results in the undesirable state of mock-modesty, but the attitude is unlikely to vanish completely if only as a reaction against modern publicity methods in other non-scientific spheres of activity such as public relations or advertising. In other countries, the reverse situation exists; a certain degree of over-statement is the accepted practice and one which must be followed to obtain due recognition.

In either case, there is no disgrace implied in ending with a sort of open-minded conclusion; no-one can be expected to solve the riddle of even one small part of the universe in a single paper.

Acknowledgements

These usually present little problems, ranging as they do from the strictly formal to the personal appreciation. One difficulty does arise, however, when making the decision whether to invite co-authorship or simply to thank the collaborator for his or her help in this section. In many cases it can be very tactful to invite someone to become a co-author, even if the bulk of the work is still done by the first-named writer. At other times a senior scientist can often give valuable help to a junior in this way, but it is perhaps best to avoid a multiplicity of authors' names. Certainly a mention of more than three names in the title of a relatively short paper is liable to invite suspicion, even though the circumstances may warrant their inclusion.

References

The choice of references to previously published works on the same subject is not easy. It is most difficult when the author is one of the very few people working in the subject, if not indeed the only one. Under these circumstances a constant repetition of his own name in the list of references suggests a degree of chauvinism, whereas in fact he is often only too willing to mention other works if only he could find them.

In most cases, it is probably advisable only to include in the list the specific papers which have been referred to in the text, but this at times can be unduly restrictive. The situation to avoid is a long list of references, included uncritically without reference to content or standard, which on the surface looks imposing, but in practice may have the reverse effect. For a review paper, however, every attempt should be made to compile a comprehensive list.

The standard format of the publisher or editor must be adhered to, and it is often helpful to include the language in which the reference paper is written, especially if the title itself has been translated.

Tables

The great advantage of a written paper, in contrast to a spoken presentation, is that far more details than on any slide or visual aid can be included in the tables. Although they may be submitted on separate sheets at the end of a script, they will be generally inserted within the text as close as possible to the places indicated by the author. Difficulties of paging a journal often make a precise placing impossible, but they will be close enough to the relevant paragraphs to allow the reader to switch his eyes from the text to the data.

The author should do his best to construct the size of the table so that it fits the dimensions of the printed page in the journal of his choice. This helps the editors and sub-editors enormously and saves the author from frustration at the proof stage.

For the purposes of permanent record it is sometimes desirable to include fairly large amounts of data in the form of an annex or appendix. This is to be recommended when the details are not essential for the reading and understanding of the paper, but are worthy of printing on their own account. A continuous series of records generally finds its way into some form of regular data publication, but a valuable set of observations which are only taken over a limited period can be lost to other workers unless they reach print in this way.

Figure captions

Diagrams and figures should be used sparingly in a written paper, if only for the reason that they take up a lot of space and cost money to reproduce. In the verbal presentation of a subject, a large number of slides may be used; in writting up the same topic for publication, a far smaller number of illustrations is advisable.

The captions should be as self-explanatory as possible, and should not err on the side of brevity if in so doing a degree of ambiguity or uncertainty is introduced. The diagrams themselves can be in greater detail than in the case of a lecture slide, but even so a very intricate and overloaded diagram defeats its own purpose. The best diagram is one which makes a dramatic point or renders more intelligible a mass of indigestible figures, for it must be acknowledged that not every reader treats numbers as a familiar language.

Sometimes it is possible to include photographs within the published paper. These are most desirable when illustrating the nature and exposure of a measuring instrument, which is almost impossible to describe accurately in words. They are also helpful in illustrating, for example, a plant response to various treatments. The use of photographs for sake of photography is to be deprecated, but they can often provide evidence that cannot be communicated in any other satisfactory way.

Referees

Even the best written paper will stand or fall according to the temper of the referees the editor calls in to help him. The problem is not made easier by the fact that referees often disagree completely amongst themselves, leaving the editor and author in a state of mixed annoyance and bewilderment.

Very few referees are actively hostile to the potential author. If they are, there is only one thing to do, and that is to retreat with as much grace as possible. It is futile to try and convince them or placate them, and the writer must simply try elsewhere or bide his time with as much patience as he can muster and return later when the climate of criticism is more favourable.

Most referees are helpful, although they may not appear to be so at the time. The truly helpful referee can be of great assistance to the young writer and his advice should be gratefully accepted, even though constant rewriting of a paper is hardly an enjoyable pastime. The real troubles lie with the apparently helpful, who seem to spend their time asking the author to rewrite in the style of the referee or insisting on the inclusion of details which most people would consider unnecessary. How patient it is necessary to be with such individuals depends on the author's age and standing, the repute of the journal concerned, and the possibility of finding an alternative outlet for the paper under consideration. When young it is advisable to imitate the wise bamboo and bow to the wind. When older and when hope, which was so invigorating a breakfast has become so unsatisfactory a supper, it simply is not worth the trouble to sacrifice your own ideas of quality for someone else's which may or may not be superior. Everyone thinks that his own children are beyond reproach, or at any rate resents their deficiencies being criticized, but one of the compensations of maturity is the ability to distinguish between occasions when major principles are at stake and when it is ridiculous to waste time and temper on minor points of issue.

In short, therefore, do not be discouraged if someone, at a referee's behest, asks you to change some parts of a submitted paper; despite your faith in your own ability, it may be an improvement. On the other hand, do not spend more time than you can spare in constant revisions, unless the stakes are high enough to justify the labour. One final consoling thought, just as more criminals escape conviction than innocent people are punished, far more poor papers reach publication than brilliant papers are rejected.

Proof reading

Always seek help in proof reading, the author cannot do it by himself. The reason for this is that his eyes and brain will tend to read what they are expecting to read and not what is actually printed. In any case tables of data can rarely be checked singlehanded.

Don't change your mind in the middle of a proof and suddenly wish you had written a sentence in a different manner. Alterations cost money and alienate editors. Attend to proofs immediately, even if you appear to be the only one involved who has to operate at instant speed, with long delays from all others concerned.

Popular writing

Writing for a non-scientific public is a different art altogether and the only thing it has in common with the writing for scientific journals is the need to give maximum consideration to the type of publication and the type of reader. If you have a gift for this style of writing, by all means use it, the results can be very remunerative to you and helpful to others.

Book writing

Don't. The writing of 2,000 words is comparatively easy; 10,000 words just about possible without undue pain; the writing of a book is sheer un-diluted hard work.

Annotated List of Technical Notes on Subjects of Agricultural Meteorology, Published by the World Meteorological Organization, Geneva

> "They lard their lean books with the fat of
> other's works."
> ROBERT BURTON

> "When I am dead, I hope it may be said,
> His sins were scarlet, but his books were read."
> HILAIRE BELLOC

Technical Note No.10   The forecasting from weather data of potato blight and other plant diseases and pests (48 pp.)

Author   P.M.A. Bourke

This is an early report on disease forecasting, dealing mainly with the methods used to forecast the incidence of potato blight in The Netherlands, the United Kingdom, Norway, Ireland, U.S.A. and Canada.

The second part of the publication treats briefly the following diseases and pests: onion smut, potato root eelworm, brown rust of wheat, vine mildew, rice stem borer, cacao black pod, European corn (maize) borer, white pine blister rust, pale western cutworm, tobacco downy mildew. 116 references are given, all earlier than 1955.

Technical Note No.20   The climatological investigation of soil temperature (18 pp.)

Author   M.L. Blanc

Although this report was published as long ago as 1958, it remains a good source for basic information regarding the problems of soil temperature measurement. It first discusses the physical principles of heat exchange from air to soil, soil to air, and within the soil. Dealing briefly with biological significances and with soil moisture effects, it concludes with an account of the observational techniques and a short bibliography.

Technical Note No.21   Measurement of evaporation, humidity in the biosphere, and soil moisture (49 pp.)

Author   N.E. Rider

The first section of this report, published in 1958, discusses the suitability of instruments for determining evaporation and evapotranspiration, including the water budget approach, the energy budget approach, the turbulent transfer approach and the use of evaporation tanks, pans, atmometers, evapotranspirometers and lysimeters. The appendix to this section presents the reasons why the Class 'A' pan was selected as a standard measurement in the IGY programme.

The second section deals with the suitability of instruments for the measurement of humidity, including wet- and dry-bulb thermometers, psychrometers (Assmann, sling, and thermo-electric), resistance thermometers and thermistors, electro-chemical and mechanical hygrometers, summarizing their advantages and limitations.

The third section concerns soil moisture measurements, including gravimetric methods, tensiometers, electrical and thermal methods, and the neutron scattering method, again specifying the difficulties involved. The conclusions reached in this section that "the development of a suitable apparatus looks remote at present" and "there appears to be no substitute for the laborious task of oven-drying numerous samples", would still be valid today.

Each of these three types of measurement are difficult to make and no major advances have been made in the 15 years since this report was published; the difficulties are still largely unsurmounted.

Technical Note No.32   Meteorological service for aircraft employed in agriculture and forestry (32 pp.)

Authors   P.M.A. Bourke; H.T. Ashton; M.A. Huberman; D.B. Lean; W.J. Maan; A.H. Nagle

Published in 1960, this report discusses the scope of agricultural aviation, including anti-locust flights, applications of forest pesticides, weed killing, fungicide spraying, forest fire protection systems and surveys.

The second chapter outlines the meteorological aspects of this type of work both in respect of the flying problems and the operational efficiency. Then follows summaries of the needs for specialized meteorological knowledge, additional instruments and observations, and designs for a routine information service; some 60 references are quoted.

Technical Note No.41   Climatic aspects of the possible establishment of the Japanese Beetle in Europe (9 pp.)

Author   P.M.A. Bourke

This problem arose from the spread of this pest from Japan to U.S.A. To assess the possibility of further spread, the author records the present geographical distribution of the beatle and the meteorological factors affecting its life cycle, thereby deducing an approximate meteorological model for a suitable environment.

The climate of Europe was then examined in order to delineate the areas at risk. Published in 1961, this report can be recommended as a prototype for this kind of problem.

Technical Note No.42   Forecasting for forest fire services (56 pp.)

Authors   J.A. Turner; J.W. Lillywhite; Z. Pieslak

After explaining the problem and defining the specialized terms used, this report explains the fire—weather relationships and the influences of meteorological conditions on the occurrence, spread and control of forest fires.

The methods of fire—weather investigation are then discussed, followed by a description of the existing methods for providing special forecast services to forest authorities with detailed examples from several countries.

This report was published in 1961 and contains many references to published papers. Subsequent research has developed more fire-danger indices and more recent publications should be referred to for the present status of the problem, but the basic principles remain.

Technical Note No.43   Meteorological factors influencing the transport and removal
                              of radio-active debris (171 pp.)

Author   W. Bleeker (editor)

Twelve authors have contributed to this publication. It can be regarded as a source volume for information concerning problems of the airborne transfer of small pollutant particles such as pollen or pathogens.

A more direct agricultural implication is the danger of contamination of food crops or animal products such as milk through the eating of polluted fodder, but this is not dealt with in the report (published in 1961).

Technical Note No.51   Protection against frost damage (62 pp.)

Authors   M.L. Blanc; H. Geslin; I.A. Holzberg; B. Mason

This report, published in 1963, starts by outlining the physical processes which determine the incidence of frost. It then discusses the warning and advisory services, both climatological and synoptic, which can be provided to the farmer or grower.

The section on protection deals first with site selection, choice of growing season and the questions of varietal influence. It then discusses helpful cultural practices involving soil and plant management. Finally, it reviews the active methods of combating frost, cover, artificial fogs, wind machines, water sprinkling and the supply of additional heat.

A good annotated bibliography is included in this report, which deals with an important subject in which work is always in progress. It is doubtful, however, if any major significant advance has been made in more recent years, although new methods are constantly under trial; probably the only significant addition is the attempted use of plastics and plastic foam.

Technical Note No.53   The effect of weather and climate upon the keeping quality of
                              fruit (180 pp.)

Authors   G.D.B. de Villiers; D.S. Brown; R.G. Tomkins; G.C. Green

This extensive and detailed report contains a wealth of useful information covering all aspects of this important economic problem. The influential factors can be regarded as environmental (climate, weather and soil), cultural, or operational (post-harvest handling, treatment and storage). After detailing the non-meteorological effects, the authors consider the effects of weather and climate on the keeping quality of deciduous fruits (apples, pears, peaches, plums, apricots, cherries, strawberries, grapes) and then deal with tropical and subtropical fruits (dates, figs) and citrus. Practical guidance is given to the grower on the storage and disposal of the crop.

The second and third parts of this publication deal with the banana and pineapple, respectively, in a most comprehensive manner.

A large amount of information is included concerning the incidence of various pests and diseases, and a very extensive bibliography is also included.

This is a unique document on this subject and can be recommended for both quantity and quality. Published in 1963, and unlike some of the fruits therein discussed, it keeps very well.

Technical Note No.54   Meteorology and the migration of desert locusts (115 pp.)

Author   R.C. Rainey

This, subtitled "Applications of synoptic meteorology in locust control" and also referred to as *Anti-locust Memoir No.7* of the Anti-locust Research Centre, London, is a major contribution to a very important problem.

After outlining the relevant aspects of the biology and behaviour of the locust, the report discusses the displacements of individual swarms in relation to meso-scale weather factors. It then gives a detailed account of the geographical distribution and movements of locusts, during 1954—55, in relation to the concurrent synoptic meteorology, covering large areas of Africa and southwest Asia.

Several of the recommendations put forward in this report have now been put into effect with very encouraging results. Published in 1963, the report contains many excellent diagrams and charts and an extensive bibliography.

Technical Note No.55   The influence of weather conditions on the occurrence of apple scab (including a report on instruments recording the leaf wetness period) (41 pp.)

Authors   _   J.J. Post; C.C. Allison; H. Burckhardt; T.F. Preece

Appendix Authors   F. Schnelle; L.P. Smith; J.R. Wallin

This report starts with a short historical review and some general remarks on the biology of apple scab (*Venturia inaequalis*), following with a summary of the investigations concerning this disease which had been carried out in Belgium, Finland, France, Germany, The Netherlands, South Africa, Switzerland, the United Kingdom and the United States of America. The warning systems existent in ten countries are then referred to, and 34 references are quoted.

The Appendix deals in some detail with 16 instruments which have been designed to measure the duration of leaf wetness; where available the manufacturer's name and address are given. Comments are made on the behaviour of many of the instruments, and the basic design requirements are specified.

The report was published in 1963, but has not been superceded, nor has the list of instruments been supplemented at any later data, although newly designed models have occasionally been referred to in various published papers.

Technical Note No.59   Windbreaks and shelterbelts (188 pp.)

Authors   J. van Eimern; R. Karschon; L.A. Razumova; G.W. Robertson

This again is a major contribution on an important subject. It deals in extensive detail with the influence of shelterbelts on air flow, heat balance, water balance, soil erosion, and air quality and then describes the resultant effects on plants, livestock, and buildings.

The first appendix contains 74 figures and illustrations; the second a very extensive bibliography.

Published in 1964, it can be strongly recommended as comprehensive review of the problem. Even a decade later, it would be difficult to suggest any major modifications or additions.

Technical Note No.65   A survey of human biometeorology (113 pp.)

Authors   F. Sargent and S.W. Tromp (editors)

Although this publication deals chiefly with human problems, Chapter VII devotes eleven pages to consideration of the influence of weather and climate on farm animals and on insects affecting the health of man.

The following topics are briefly dealt with: optimum thermal environments, the reactions to heat and cold and modifications thereto caused by age, species, type, acclimatisation and altitude.

The entomological aspects are discussed in general terms, pointing out some effects of temperature, humidity, wind, radiation and pressure.

This report was published in 1964; ten pages of references are included, some of which refer to research in agricultural meteorology.

Technical Note No.83   Measurement and estimation of evaporation and evapotranspiration (121 pp.)

Authors   M. Gangopadhyaya; G.E. Harbeck; T.J. Nordenson; M.H. Omar; V.A. Uryvaev

This report, published in 1966, gives details of the various attempts that have been made to measure or calculate these parameters, but is not designed as a critical survey.

The chapter dealing with direct measurements includes details of atmometers, pans and tanks, contains useful advice regarding their exposure and use, and quotes some results on various comparison tests in a number of countries. A further chapter deals with evapotranspirometers and lysimeters, including diagrams showing their construction.

Details of methods of estimation include the water-budget methods and the energy-budget methods, and quotes a variety of relevant formulae proposed by various authorities. Short bibliographies throughout the book enable the reader to refer back to original publications.

Technical Note No.85   Accuracy of measurements by pyrheliometers (130 pp.)

Authors   A. Angstom et al.

This report of a meeting held in Brussels in 1966, contains twelve papers by leading experts in radiometry.

It is worth reading to understand the difficulties that arise in this type of measurement.

Technical Note No.96   Air pollutants, meteorology and plant injury (73 pp.)

Authors   E.I. Mukammal; C.S. Brandt; R. Neuwirth; D.H. Pack; W.C. Swinbank

After a general review of the problem, this report deals with the sources and chemistry of the air contaminants, followed by descriptions of the injuries to plants due to pollution effects on their physiology, and the symptoms of such injuries.

The meteorological processes affecting the emission, spread and concentration of pollutants are then dealt with in some detail, including consideration of source configurations, vertical and horizontal turbulence and diffusion, and transport through the atmosphere, leading to their application to problems of pollution potential and concentrations. Further sections discuss sampling, instrumentation and forecast systems, finally suggesting some measures of control.

This report, published in 1968, contains over 170 references, but with the increased awareness of the dangers of increasing atmospheric pollution, a large number of more recent papers are now available. The basic principles outlined in this publication remain a valuable contribution to an important problem.

Technical Note No.97   Practical soil moisture problems in agriculture (69 pp.)

Authors   W. Baier; J.J. Doyle; M. Gangopadhyaya; L.A. Razumova; G. Stanhill; E.J. Winter

This work is divided into four chapters, the first of which contains a list of published reviews on the problems of the soil moisture balance. Each entry includes a brief summary of the contents. Some indication of the volume of published material on this subject can be gained from the fact that in this short list, the reviews include a grand total of over 3,000 references.

The second chapter discusses the various methods of measurement of soil moisture, including direct weighing, electrical resistance, tensiometer and neutron-scattering methods. It includes comments on the usefulness and validity of each type of method.

The third chapter, almost half the publication, surveys the many attempts that are made to measure evaporation and transpiration for irrigation purposes, and the various methods of calculating such values from meteorological data. This survey is fairly comprehensive and contains important comments on possible difficulties and sources of error, although it does not attempt any full critical comparisons. The comments throughout are made with the practical applications in mind; field accuracy rather than scientific precision is the preferred criterion.

A short fourth chapter selects subjects in which further research is needed, including the identification of the best soil moisture regime for various crops, comparison tests of the accuracy of methods of measurement, better instruments, the special problems of irrigation under very dry climatic conditions and the effects of high water tables.

There are two appendices. The first gives details of the methods used by the U.S.S.R. Hydrometeorological Service for routine measurement of soil water content. The second reviews methods in use in various parts of the world under the geographical headings of Africa, America, Europe, Southeast Asia and Australia, Russia and Eastern Europe. In addition there is a short list of 62 references.

The report was published in 1968. Much of the material and especially the informed comments, will remain valid for many years, despite the fact that there is considerable activity in this important subject.

Technical Note No.99   Meteorological factors affecting the epidemiology of wheat rusts (143 pp.)

Authors   W.H. Hogg; C.E. Hounam; A.K. Mallik; J.C. Zadoks

This extensive report is one of the important basic documents concerning the meteorology of the rust diseases of wheat. It first deals with a general description of black, brown and yellow rusts, including a description of the relationship between the weather factors and the physiological processes of the diseases.

Then follows a valuable survey over all parts of the world of the information concerning the occurrence and severity of the diseases. The questions of spore transport and the epidemiology of rusts is then dealt with in considerable detail, and finally the possibilities of warning forecasts is discussed.

Three appendices contain, respectively, a glossary of phytopathological terms, a glossary of meteorological terms and an extensive bibliography.

The authors sought help from meteorologists and pathologists in over 25 countries and produced a report, published in 1969, which could serve as a model for any similar investigation. It is strongly recommended not only for the information it contains on this particular set of diseases, but also as an indication of how a multi-disciplinary problem should be treated.

Technical Note No.101   Meteorology and grain storage (65 pp.)

Author   C.V. Smith

The opening chapter of this report deals with the biological characteristics of the grains; viability, fungi, insects and mites. This is followed by a discussion of the physical relationships within the bulk storage; water, heat and air movement. The pre-storage treatment is then considered, including artificial drying; a further chapter on storage management includes details of ventilation and cooling.

The final chapters discuss the instrumentation necessary for measuring storage conditions and the role of meteorological help in reducing storage losses.

The report, published in 1969, includes an extensive bibliography and a series of helpful diagrams.

Technical Note No.107   Meteorological observations in animal experiments (37 pp.)

Author   C.V. Smith

In animal production, as in plant production, one of the main research problems concerning the agricultural meteorologist is that of heat balance, which also involves questions of the moisture balance and latent heat. The main physical processes involved are radiation, convection, conduction and evaporation.

The first section of this report deals with outdoor experiments, dealing with the physiological relevance of various factors, and specifying the requirements for measurements of energy exchange, temperature, humidity, air movement, precipitation, evaporation, pressure, air quality, irrigation and certain other derived or combined parameters; there is a short reference to problems of animal disease.

The second section is concerned with indoor experiments, considering ambient temperatures and humidities, together with air movement and air changes within the structures. Helpful advice is given regarding the performance of various types of instrument, the use of tracers for determining airflow pattern, and calibration and control procedures.

The report, published in 1970, concludes with some useful recommendations and a short bibliography. The amount of work of this nature is not extensive and the advice in this publication should enable a research worker to overcome many of the difficulties involved.

Technical Note No.113   Weather and animal diseases (49 pp.)

Author   L.P. Smith

The opening chapters of this report discuss the types of problem, methods of analysis and previous reviews of the subject; the most recent work is then considered in greater detail.

The spread of fowl pest and foot and mouth disease is given as an example of the windborne spread of infections and mention is made of airborne pollution. Stress conditions are discussed with reference to both environmental conditions and nutritional deficiencies consequent on weather conditions.

Parasitic diseases such as fascioliasis (liver-fluke), nematodiriasis, parasitic gastroenteritis, parasitic bronchitis, red-water fever and leptospirosis are then dealt with, followed by fungal diseases such as facial eczema, mycotic abortion and mycotic dermatitis. There is a brief reference to the effect of meteorological conditions on fecundity and fertility.

The appendices include more detailed analyses of seven subjects of recent investigation, and there is a bibliography of over 120 references.

This report was published in 1970 and presents a unique collection of work in this field of research.

Technical Note No.118   Protection of plants against adverse weather (64 pp.)

Authors   G.W. Hurst and R.P. Rumney

This is an extensive review of a large number of published papers, all dealing with crop protection or climate amelioration. The elements are dealt with in turn; temperature, sunshine and light, precipitation and irrigation, wind, and derived effects such as erosion and salt-spray damage.

Published in 1971, it includes a large annotated bibliography and a number of illustrations and diagrams.

It can be recommended as a reference volume on a wide variety of associated topics; the coverage of published papers up to 1968 is almost complete.

Technical Note No.119   The application of micrometeorology to agricultural problems
                             (74 pp)

Authors   A. Baumgartner; M. Budyko; E. Inoue; E.R. Lemon; J.L. Monteith; R.O. Slatyer; G. Stanhill; L.P. Smith (editor)

This analysis is divided into three parts. The first part, under the six main headings of radiation, momentum, heat, water, carbon-dioxide, and finally all other chemical and biological matter, is concerned with the identification of the scientific processes which determine their existence and transfer throughout the biosphere. It gives a brief résumé of each process and classifies them with respect to the type of science, soil, atmospheric, plant or animal, which is chiefly concerned with their investigation. It also includes the processes of measurement and the responses of biological material to physical conditions. The review is comprehensive; it explains the relative importance of each component and indicates to some extent the degree of knowledge therein at present available; it does not enter into detailed expositions.

The second part classifies the types of practical problems in agro-meteorology and agro-climatology and specifies which of the processes nominated in the earlier section are relevant to each problem, again attempting to identify those of the greatest importance, thus giving some idea of the research priorities in regard to operational demand. A further summary is included which indicates the roles of the four main groups of scientific disciplines.

The third part contains a short review of the present state of the subject and makes suggestions as to its future progress, defining its role and responsibilities towards other research workers both within and on the fringe of its own field, and to the application of its findings to advisory, educational and training programmes.

There are two appendices; one gives the titles and addresses of organizations engaged in micrometeorology and research, with some indication of their special interests; the other is a limited bibliography of the more important of recently published papers.

This report was published in 1972 and is strongly recommended as being the first attempt to present a comprehensive view of the subject and its potential.

Technical Note No.122   Some environmental problems of livestock housing (71 pp.)

Author   C.V. Smith

This report considers the living conditions of housed animals in the light of meteorological physics, dealing chiefly with the climates of the temperate zones.

After pointing out that bigger and more expensive buildings are not necessarily an improvement on older systems, the thermal environment and heat balance problems are discussed. The quality of the air and the living conditions in respect to pollutants is then considered in some detail, including the impact of airborne material such as dust, pollen, pathogens and disease vectors, and also the chemical composition of the air.

Air movements around and inside the buildings, natural and artificial ventilation are then fully discussed together with the resultant effects on the heat balance. Then follows a chapter on instrumentation and monitoring of the internal climates and a discussion of management problems.

The author suggests that although the cost of achieving a high quality living environment may be considerable, the future of animal husbandry may include semi-automatic control of internal climate, provided that adequate meteorological assistance is available at the research and development stages. The most serious limitations to large systems seem to lie within the problems of health and hygiene.

Each chapter ends with a list of references and there is a general bibliography at the end.

This report was published in 1972 and is the only publication to date in which this important subject is treated from the point of view of the meteorologist. The number of mistakes that have been made in the design of modern buildings for animals underline the necessity for information of this nature.

APPENDIX II

Annotated Bibliographies

(1) Issued by Commonwealth Bureau of Pastures and Field Crops

(a) *Field Crop Abstracts:*
No. 1008	Water relations of the rice plant (70 refs.; 1960—65)
No. 1047	Effect of water stress on the growth and development of ground nuts (21 refs.; 1957—67)
No. 1071	Sprinkler irrigation of rice (8 refs.; 1951—57)
No. 1140	Effect of temperature on the germination of oats (10 refs.; 1948—68)
No. 1182(1)	Water requirements of crops — rice (106 refs.; 1960—70)
No. 1182(2)	— wheat (118 refs.; 1960—69)
No. 1182(3)	— maize (85 refs.; 1965—69)
No. 1182(4)	— sorghum, millet (44 refs.; 1960—69)
No. 1182(5)	— cotton (127 refs.; 1960—69)
No. 1182(6)	— soyabeans (58 refs.; 1960—69)
No. 1182(7)	— groundnuts (31 refs.; 1960—69)
No. 1182(8)	— sunflower (20 refs.; 1960—69)
No. 1182(9)	— methods for determining (95 refs.; 1960—69)
No. 1230	Windbreaks in the tropics (26 refs.; 1947—70)
No. 1264	Irrigation of potatoes (135 refs.; 1953—71)
No. 1278	Anti-transpirants (25 refs.; 1949—71)
No. 1309	Effect of light on the growth and yield of potatoes (74 refs.; 1962—72)
No. 1316	Flooding effects on temperature crops (19 refs.; 1960—72)
No. 1340	Crop growth models (16 refs.; 1968—72)

(b) *Herbage Abstracts:*
No. 1199	Haymaking in the tropics (111 refs.; 1950—70)
No. 1272	Field curing of hay in relation to environmental conditions (64 refs.; 1961—71)
No. 1314	Flooding effects on herbage (86 refs.; 1949—72)
No. 1330	Sorghum-effect of environmental conditions (215 refs.; 1963—72)

(2) Issued by Commonwealth Bureau of Horticulture and Plantation Crops

No. 5703	Responses of tropical and sub-tropical crops to light intensity (51 refs.; 1951—69)
No. 5725	Responses of cacao to temperature (20 refs.; 1948—69)
No. 5742	Irrigation of various vegetable crops (104 refs.; 1966—70)
No. 5802	Trickle irrigation (36 refs.; 1955—70)
No. 5929	Date and duration of flower-bud differentiation in apples and pears (37 refs.; 1955—70)
No. 5981	Air-supported plastic greenhouse structures (28 refs.; 1961—71)
No. 6010	Tree crop responses to wind (25 refs.; 1961—71)
No. 6025	Wind damage and lodging in sugar cane (27 refs.; 1953—70)
No. 6034	Responses of vegetables to artificial and supplementary light (115 refs.; 1964—72)
No. 6080	Controlled environmental facilities (51 refs.; 1968—72)

(3) Issued by the Commonwealth Bureau of Soils

No. 884   Evaporimetry for scheduling irrigation (61 refs.; 1951—64)
No. 1015   Soil losses on steep slopes (48 refs.; 1940—65)
No. 1032   Methods of measuring evapotranspiration (79 refs.; 1956—65)
No. 1083   Seasonal soil variations (90 refs.; 1962—66)
No. 1084   The effect of radiation on plants (51 refs.; 1956—65)
No. 1097   The effect of forest shelter belts on soil and plants (35 refs.; 1955—66)
No. 1106   Reduction of evaporation from the soil (58 refs.; 1956—66)
No. 1109   Effect of flooding or high soil-moisture content on cereals and grasses (73 refs.; 1954—67)
No. 1131   Water table in relation to growth and nutrition of plants and drainage (59 refs.; 1956—67)
No. 1187   Heat flow in soil (46 refs.; 1948—67)
No. 1191   The plant-nutrient content of rain water (48 refs.; 1962—67)
No. 1203   Wind erosion and its control (134 refs.; 1950—67)
No. 1237   Effects of moisture and drying on soil microflora (47 refs.; 1964—68)
No. 1272   Fertilizer response and moisture supply (92 refs.; 1963—68)
No. 1276   Water relations of wheat including irrigation (108 refs., 1955—68)
No. 1377   Effect of soil temperature and moisture and light on growth of forest tree seedlings (47 refs.; 1956—69)
No. 1379   Irrigation and water relationship of temperate fruits — apple, pear, cherry, plum, apricot and peach (67 refs.; 1956—69)
No. 1380   Ditto — citrus, orange, grapefruit, lemon, olive, mulberry, bushfruits, currants, grape vine, raspberry, blueberry, strawberry (87 refs.; 1956—69)
No. 1393   Water relationships and dry matter of grassland herbage as affected by N supply (91 refs.; 1965—69)
No. 1402   Rate and measurement of soil erosion (52 refs.; 1963—69)
No. 1421   Water relations of potato, including irrigation (86 refs.; 1956—69)
No. 1430   Some references to soil-water/vegetation relationship in tropical and subtropical regions (68 refs.; 1965—70)
No. 1444   Plant nutrition and growth as affected by soil and root temperatures (93 refs.; 1968—70)
No. 1543   Effect of light and shade on plant nutrition and growth (67 refs.; 1966—71)
No. 1553   Some references to lysimeter and lysimetric studies (59 refs.; 1965—72)
No. 1555   Some references to air pollution of soils and plants (74 refs.; 1966—72)
No. 1577   Water in rice nutrition and growth (66 refs.; 1961—72)
No. 1586   Nutrient uptake, growth and composition of crop plants as affected by soil moisture stress (113 refs.; 1966—72)
No. 1597   Nitrate leaching loss from soil; some references to nitrate in waters (106 refs.; 1966—72)
No. 1601   Some references relating to nutrients occurring in irrigation and drainage waters (52 refs.; 1963—73)

The Bureau of Soils has also published three important *Technical Communications:*
No. 29   Soil, vegetation and climate,   by G.V. Jacks, 43 pp.
No. 49   Mulching   by G.V. Jacks, 87 pp.
No. 53   Vegetation and hydrology,   by H.L. Penman, 124 pp.

Subject Index